PRINS AND PNA TECHNOLOGIES IN CHROMOSOMAL INVESTIGATIONS

PRINS AND PNA TECHNOLOGIES IN CHROMOSOMAL INVESTIGATIONS

FRANCK PELLESTOR
EDITOR

Nova Biomedical Books
New York

For permission to use material from this book please contact us:
Telephone 631-231-7269; Fax 631-231-8175
Web Site: http://www.novapublishers.com

Library of Congress Cataloging-in-Publication Data

PRINS and PNA technologies in chromosomal investigation / Franck Pellestor (editor).
p. ; cm.
Includes bibliographical references and index.
ISBN 13 978-1-60021-413-4
ISBN 10 1-60021-413-4
1. Nucleic acid probes. 2. In situ hybridization. I. Pellestor, Franck.
DNLM: 1. Primed In Situ Labeling. 2. DNA Probes. 3. In Situ Hybridization, Fluorescence. 4. Peptide Nucleic Acids--genetics. QZ 52 P957 2006]
QP624.5.D73P75 2006
572.8'4--dc22 2006025942

Published by Nova Science Publishers, Inc. ✢ New York

Contents

Preface

Advances in molecular biotechnology have greatly improved the sensitivity and the efficiency of methods utilised for genetic investigations and diagnosis. In the domain of chromosome analysis, the introduction of molecular techniques has led to the development of a new approach, called Molecular Cytogenetics, which has surpassed previously available techniques to become a foremost biological method. The fluorescence in situ hybridization (FISH) is quickly became the standard technique for in situ chromosomal investigations, as illustrated by its large variety of applications in research and diagnosis. However, during the last decade, alternative methods to FISH have been introduced and have shown to be valuable in detecting chromosomes and quantifying chromosomal abnormalities. These alternative procedures are the Primed IN Situ (PRINS) labelling and the Peptide Nucleic Acid (PNA) probes. The two procedures present several advantages for the in situ detection of nucleic acid sequences, such as the small size of PNA probes and PRINS primers, or the fast kinetics of PRINS and PNA labelling reactions, that make them very attractive for a number of cytogenetic purposes.

The purpose of this book is to provide for the first time in one place, a valuable introduction and overview of the principles and the applications of these 2 alternative approaches in the field of the Molecular Cytogenetics.

The first seven chapters describe PRINS and its developments applied to chromosomes. The use of PRINS is described in various applications in human including characterization of unique chromosome sequences, assessment of aneuploidy in human gametes, detection of fetal cells, use in concert with immunohistology, as well as in plant cells. The eighth chapter addresses a particular variation of the PRINS, the padlock technique, based on the use of circularizable oligonucleotide probes, which offer a high efficiency for in situ detection of single nucleotide polymorphisms.

The final section of the book covers PNA technology and its adaptation to human cytogenetics, with four chapters illustrating the potential usefulness of PNA for chromosome investigations. These chapters described specific research and diagnostic applications of PNA to chromosomes in mitotic and non-dividing cells.

This volume contains contributions from experts in the field, making it a valuable reference source. It should constitute an excellent introduction for any scientist, cytogeneticist and cytologist as well as students in genetic programmes, to the use of PRINS and PNA technologies in chromosome analysis.

In: PRINS and PNA Technologies in Chromosomal Investigations ISBN 1-60021-413-4
Editor: Franck Pellestor, pp. 1-12

Chapter 1

Principles of the PRINS Technique

Franck Pellestor *

Department of Reproductive Biology B,
Arnaud de Villeneuve Hospital, Montpellier, France
Inserm, U 847, Montpellier, France

Abstract

The advent of molecular genetic techniques has brought forth new procedures for in situ chromosomal analysis. One of these techniques is the primed in situ (PRINS) labelling which constitutes a fast and efficient alternative to conventional fluorescence in situ hybridization (FISH) for chromosomal detection. Based on the use of chromosome-specific primers, the PRINS method combines the high sensitivity of the Polymerase Chain reaction (PCR) with the cytological localization of DNA sequences. Using this approach, numerous cytogenetic applications have been successfully developed on various types of cells. With the development of rapid protocols producing reliable and reproducible results, the PRINS procedure has an enormous potential for cytogenetic investigations.

Keywords: PRINS, chromosome, aneuploidy, in situ labelling

Introduction

The molecular cytogenetics has become an important tool for chromosome research and cytogenetic diagnosis, which has quickly surpassed previously available techniques to become a foremost biological method. Thus, the fluorescence in situ hybridization (FISH) is now a well-established technique for studying chromosomal aneuploidy and rearrangements,

* Address for correspondence: Dr. Franck Pellestor, Department of Reproductive Biology B, Hôpital Arnaud de Villeneuve 371, avenue du Doyen Gaston Giraud, 34295 Montpellier Cedex 5, France. Tel: 33 (0)4 67 33 64 04; E-mail: Franck.Pellestor@igh.cnrs.fr

gene localization and genomic organization. An interesting alternative approach to FISH technique for chromosomal screening is the primed in situ labelling (PRINS) procedure.

Both FISH and PRINS were developed around the same period [1, 2]. Because it is relatively easy to implement and perform, and because a large variety of DNA probes and in situ hybridization kits are commercially availability, FISH has rapidly found increasing applications in cytogenetics and the technique is now the method of choice for chromosomal investigations in both medical institutions and research centres [3, 4]. Compared to such a well-established technology, the PRINS procedure might appear of minor interest and of limited usefulness. However, the PRINS reaction presents several advantages such as rapidity, high sensitivity and specificity, and these make PRINS technique an attractive complement to FISH for in situ aneuploidy detection and study of subtle chromosomal rearrangements.

Principles and Methods of PRINS Labelling

The PRINS reaction combines the high sensitivity of the PCR with the cytological localization of specific DNA sequences. The key to the PRINS technique is the use of unlabeled, short and specific oligonucleotides. Oligonucleotides are annealed in situ to complementary DNA targets on denaturated chromosome spreads, nuclei or tissue section and then act as primers for chain elongation catalyzed by a Taq DNA polymerase in the presence of free nucleotides. The visualization of generated fragments results from the incorporation of one labelled nucleotide (Figure 1).

In their original report, Koch et al [2] reported the in situ labelling of consensus centromeric repeated sequences. Subsequent studies identified oligonucleotides useful in identifying each type of repeat DNA sequences spreading over the human genome, i.e. satellite, telomeric, or Alu DNA sequences [5]. A great advantage of primers is their ability to differentiate between closely related sequences. This feature has been utilized for generating chromosome-specific primers from the alpha-satellite DNA motif [6, 7]. Alpha-satellite (or alphoid) DNA is a family of tandemly repeat sequences present at the centromere of all human chromosomes. These centromeric repeats are made up of a variable number of monomeres of 171-bp in length and are organized as alpha-satellite subfamilies. The DNA sequences of momomeres slightly deviate among subfamilies and individual chromosomes. The chromosome specificity of PRINS labelling is based on the use of primers generated from these chromosome-specific alpha-satellite DNA sequences. The complementation process between the primer and its centromeric target will be so specific that a simple mismatch between the 3'-end of the primer and the genomic sequence will prevent initiation of the *in situ* elongation by the Taq DNA polymerase. Thus, it has been possible to define specific alpha-satellite primers for most of the human chromosomes, involving some chromosomes undistinguishable by conventional FISH with centromeric probes. This is the case with chromosomes 13 and 21, which share 99.7% homology in their alpha-satellite DNA sequences [8]. To date, specific primers have been defined for 20 human chromosomes (Table 1). Efficient primers have also been defined for in situ labelling of telomeres and studying variant telomeric repeats [9, 10].

Table 1. Characteristics of the principal PRINS primers used in human cytogenetics

Name	Locus	Chromosome	Sequence	Annealing	Optimal
Alu S	alu	all	5'-GCCACTGCACTCCAGCCTGGG-3'	55	150
Alu R	alu	all	5'-GCCTCCCAACGTGCTGGGATTACAG-3'	55	150
Cons	αsat	all	5'-TCTTTTTGTAGAATCTGCAAGTGGATA-3'	54	150
J52	sat III	1	5'-ATTCCATTAGATGATGACCCCTTTCAT-3'	61	100
1c	αsat	1	5'-ATTCCATTAGATGATGACCCCTTTCAT-3'	61	100
2c	αsat	2	5'-CTGTTCAACACTGTG-3'	71	150
3c	αsat	3	5'-TGAGTTGAACACACACGTAC-3'	66	150
5c	αsat	5	5'- TTCTGTCTAGCCTTACAGGAAAA-3'	70	150
7c	αsat	7	5'-AGCGATTTGAGGACAATTGC-3'	56	100
8c	αsat	8	5'-CTATCAATAGAAATGTTCAGCACAGTT-3'	67	150
9c	sat III	9	5'-AATCAACCCGAGTGCAATC-3'	56	150
10c	αsat	10	5'-ACTGGAACGCACAGATGACAAAGC-3'	63	200
11c	αsat	11	5'GAGGGGTTTCAGAGCTGCT-3'	65	200
12c	αsat	12	5'-GTTCAATTCACAGAGTAT-3'	60	200
13A	αsat	13	5'-TGATGTGTGTACCCAGCT-3'	60	100
16c	αsat	16	5'-TTCTTTTCATACCGCATTCT-3'	53	75
17c	αsat	17	5'-AATTTCAGCTGACTAAACA-3'	51	200
18c	αsat	18	5'-ATGTGTGTCCTCAACTAAAG-3'	65	100
21A	αsat	21	5'-TGATGTGTGTACCCAGCC-3'	61	150
Xc	αsat	X	5'-GTTCAGCTCTGTGAGTGAAA-3'	68	75
D600	sat III	Y	5'-TCCATTCGATTCCATTTTTTTCGAGAA-3'	56	100

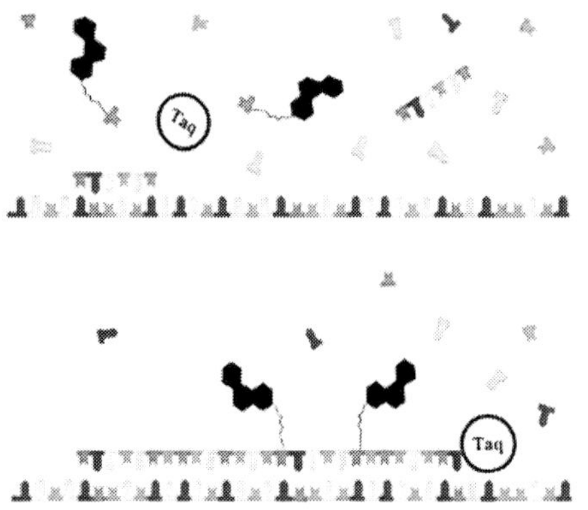

A.
Deposit of the reaction mix on denaturated preparations and specific annealing of primer to the DNA target sequence

B.
Primer extension by a Taq DNA polymerase with incorporation of the labelled nucleotide

Figure 1. Principles of the basic PRINS reaction. After annealing of oligonucleotide primers to the DNA template, the primer extension is performed by a Taq DNA polymerase in the presence of free nucleotides of which at least one is labelled. Thus, the site where the chain elongation occurs is labelled and can be visualized using fluorescence microscopy

Using an automatic DNA synthesizer, both the preparation and the purification of oligonucleotides are now fast and rather inexpensive. The length of the PRINS primers ranges from 18 to 35 nucleotides. Compared to the size of DNA repetitive probes (250 to 600 bp), this small size greatly facilitates their in situ accessibility to their genomic target sequences. This is particularly significant in cells with highly condensed nuclei, such as spermatozoa. The use of oligonucleotide primers eliminates the need for probe preparation and can also circumvent the lack of resolution of DNA probes (approximatively 5-10kb in interphase nuclei), since every known DNA sequence can be a potential primer source for PRINS reaction. As an alternative to oligonucleotides, cloned probes fragmented by restriction enzyme digestion can also been used as primers [11].

Advanced PRINS Procedures

Initially, PRINS reactions were performed either on a hotplate or a waterbath. The required changes of temperature were achieved by moving the sides as often as needed. The weakness of these methods was in the lack of stringency of primer annealing and high background. These procedures did not allow precise and durable temperature control. Indeed, although the range of Taq DNA polymerase activity is large enough, the optimization of the annealing and its stringency in order to increase specificity need accurate temperature control. The protocol has been considerably improved and simplified by the introduction of programmable temperature cyclers equipped with a flat plate block. With this equipment, the precision of temperature control may reach 0.2°C and the required temperature changes are both easy to program and rapidly carried out. The use of automatic thermocyclers allows an optimization of both annealing and extension conditions. Thus, semi automatic PRINS protocol were developed offering a high reproductibility in labelling reaction. An additional improvement was the direct use of various fluorochromes and the introduction of multiple colour PRINS reaction [7, 12] allowing the sequential labelling of several chromosomal targets on a same cell preparation. By incorporating directly fluorochrome-labelled nucleotides into newly synthetized DNA fragments, the turnaround time of the PRINS reaction was reduced to no more than a few minutes, leading to the development of ultra-fast PRINS protocols [13]. Attempts to increase sensitivity also involved the use of serial cycles of PRINS reactions for accumulating labelled copies of the target sequence at the site of synthesis [14, 15]. However, this process called Cycling PRINS could also alter the specificity of the signal and quality of the chromosome preparation [16].

In practice, a PRINS mix is prepared in a final volume of 50 µL containing the oligonucleotide (50 to 250 pM), the nucleotide mixture including a labelled dUTP (biotine, digoxigenine, fluoresceine, coumarine, rhodamine, cyanine...), the Taq polymerase buffer and 1-2 units of the enzyme. Because they are unlabeled, high amounts of primers can be used in PRINS reaction without inducing background signals. Both chemical and thermal denaturations of the preparation slide can be used. Thus, a classical PRINS procedure consists of 2 programmed steps (annealing at the specific annealing temperature of the used primer, and nucleotide chain elongation at 72°C) in case of preliminary chemical denaturation (in 70% formamide, 2X SSC at 72°C), whereas the thermic denaturation of the preparation is directly involved in the PRINS reaction, then consisting of 3 programmed steps (denaturation at 94°C, annealing and elongation). Usually, slides and coverslips are not sealed. Both the volume of the mix and the short incubation time prevent the slide from drying during the reaction. In conventional multi-target PRINS reaction, a blocking step is used between each chromosome-specific PRINS reaction to prevent the mixing of labelling and to ensure a good recognition of each targeted site. The slide is thus treated at 37°C with a dideoxynucleotides mix and Klenow enzyme which block the free 3'-ends of the elongation fragments generated by the previous PRINS reaction. Detection of the labelling sites is performed by fluorescence microscopy. Initial PRINS procedures used primers labelled with biotine or digoxigenine, and the detection of the labelling sites required immunocytochemistry procedure for detecting reporter molecules. Now, most of the PRINS reactions are performed using directly fluorochrome-labelled nucleotides, which allows the direct visualization of targeted sequences on fluorescence microscope.

Recently, a new multicolour PRINS protocol has been reported, allowing the performance of ultra-rapid detection of several chromosomes, only by mixing the different fluorochromes during the chain elongation reaction [17]. In this sequential procedure, each PRINS reaction consists of a unique short step (4-6 minutes) for annealing and elongation of each chromosome-specific primer (Fig. 2). This ingenious strategy, which eliminates the intermediate blocking reaction, greatly simplifies the PRINS protocol and facilitates its adaptation to various cytogenetic applications.

Compared with the conventional FISH technique, the PRINS method has proven to be equally accurate for detecting chromosome, quicker and also less expensive. Werner et al [18] estimated that the PRINS reaction costs approximately one-tenth as much as FISH.

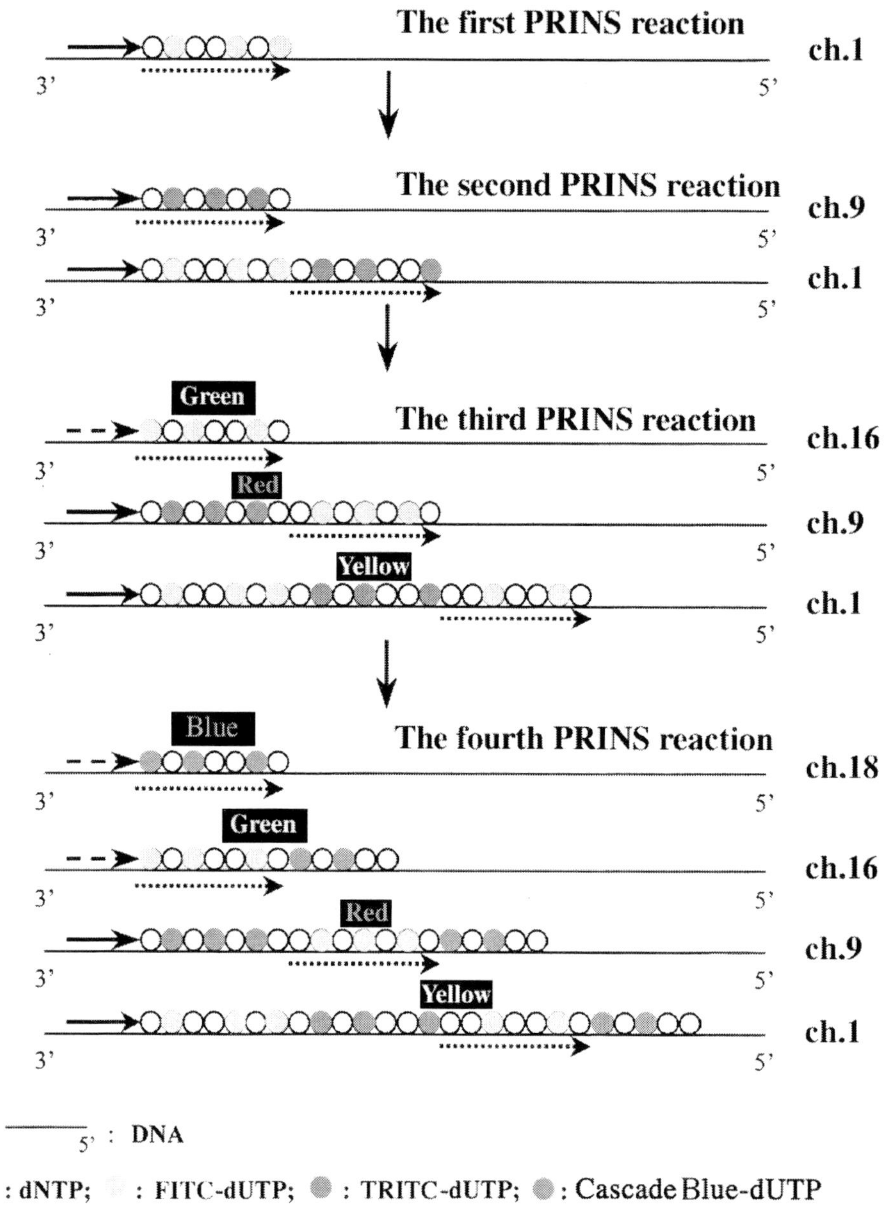

Figure 2. Principles of the multicolour PRINS reaction (modified from Yan et al. 2001). Four sequential PRINS reactions, using specific centromeric primers for 4 chromosomes and 3 distinct fluorochrome-dUTPs (FITC-11-dUTP, TRITC-6-dUTP, Cascade Blue-7-dUTP) are performed without intermediate 3'-end blocking steps. The *in situ* mixing of the colours led to the final generation of 4 distinct coloured spots (green, red, yellow and blue) allowing the specific identification of each targeted chromosome. In this illustration, the chromosome 1 is successively labelled by the in situ incorporation of FITC, TRITC, FITC and Cascade Blue, leading to a final yellow colour. The chromosome 9 is labelled by FITC, TRIC and Cascade Blue, leading to a red colour. The chromosome 16 is labelled by the incorporation of FITC and Cascade Blue, resulting in a green colour. Finally, the chromosome 18 only incorporates Cascade Blue dUTP, giving a blue colour

Applications of PRINS

The PRINS procedure combines several features, which make it very attractive for a number of cytogenetic purposes. Although PRINS was initially limited to the detection of repeat sequences, numerous applications of PRINS technology have been developed in mammalians, fish, insects and plants, demonstrating that PRINS could be easily adapted to various types of cells.

In Drosophilia, PRINS was used for gene mapping on polytene chromosomes [19], and in fish, the localization of Alu repeat and NORs was performed by PRINS [20, 21].

In plants, PRINS was used for labelling telomeres [22] and gene clusters [23]. The technique was also adapted to suspensions of bean chromosomes, in order to facilitate the flow sorting and karyotyping of similar sized chromosomes [24].

On mammalian cells, a large variety of PRINS applications have been described, and the technique has been integrated into various protocols of research. For instance, Russo et al [25] used PRINS for studying mechanisms of aneuploidy in mouse splenocytes. Because of its high sensitivity, the PRINS technique was also chosen for telomeric studies in various species [26, 27]

In humans, PRINS method has been successfully adapted for the assessment of aneuploidy in lymphocytes, amniocytes and preimplantation embryos [28, 29, 30], but also for analysis of structural aberrations such as translocations, marker chromosomes and ring chromosomes [31]. More recently, the PRINS protocol has been adapted to detect fetal cells among separated maternal nucleated cells from peripheral venous blood of pregnant women, demonstrating the efficiency of the technique for non-invasive prenatal chromosome analysis [32, 33]. As the PRINS reaction with Alu primers gives high quality R-like banding on human chromosomes, the procedure has been adapted for the cytogenetic screening of somatic hybrid cell lines [34] and identification of euchromatin in aberrant short arms of acrocentric chromosomes and small ring chromosomes [35]. Further applications have been found in tumoral cytogenetics. To date, preliminary studies involve the detection of host cells following sex-mismatched bone marrow transplantation in patients grafted for leukemia and the cytogenetic screening of tumoral cell lines [36]. The PRINS technique can also be performed in both frozen and paraffin-embedded tissue sections, or to localise RNA within cells, which opens new possibilities for cytogeneticists and pathologists [37]. All these applications point out the potential efficiency of PRINS method for diagnostic use.

Taking advantage of the high sensitivity and specificity of PRINS reaction, some laboratories have successfully adapted the technique to the in situ detection of RNA [38] and the quantitation of RNA species in cell suspensions by flow cytometry [39]. In the same way, Andersen et al [40] use a variation of PRINS (the self-PRINS) to study the methylation status of CpG islands in human chromosomes. By employing variant telomeric primers, Krejci and Koch [41, 42] clearly demonstrated the superiority of PRINS for measuring telomeric length and identifying polymorphic telomeric repeats. Studies have also shown that PRINS and FISH could be used in concert for simultaneous detection of chromosome targets [43, 44].

One of the most remarkable and efficient applications of PRINS technique has been carried out on human spermatozoa. Using PRINS technique, diploidy and disomy frequencies have been estimated for most of the human chromosomes in sperm samples from several men

[45], and studies of meiotic segregation pattern were performed in sperm samples from reciprocal translocation carriers [46]. Recently, new multicolour PRINS protocols have been successfully adapted on human spermatozoa [47] as well as on human oocytes and blastomeres [48].

The adaptation of PRINS technique to the in situ detection of unique sequences has constituted an important challenge, because this procedure could allow to rapidly map genes on chromosomes by simply using synthetic oligonucleotides derived from sequenced DNA. The advantage of the PRINS approach is that it does not rely on the possession of a cloned probe for the target. As long as the sequence of the gene is known, oligonucleotides can be synthesized for use as primers. Several teams have worked on this new application of PRINS. Preliminary results have been obtained on porcine chromosomes where unique sequences as short as 100-300 bp were localized by using pairs of specific primers [49]. More recently, the procedure has been optimized to detect single copy loci on human chromosomes. Thus, Kadandale et al [50] first localized the SRY gene in situ, and several further reports showed that PRINS could be used for gene mapping [16, 51], but also for the diagnosis of microdeletion syndromes and cancer abnormalities [52, 53]. This opens news and promising perspectives for PRINS in the field of physical mapping.

Conclusion

PRINS has emerged as a research technique, and since its introduction, the technique has undergone many stages of development to improve the sensitivity, the efficiency and the versability of the process. Numerous studies have demonstrated that PRINS is as reliable and more efficient than FISH for detecting chromosomes in metaphase and interphase nuclei. With the development of rapid and simplified protocols producing reliable and reproducible results, PRINS has become a powerful tool for cytogenetic investigations and diagnosis. During the last years, PRINS has progressed from the labelling of repeat sequences to the use for the detection of unique sequences. New innovative variations to PRINS such as padlock probes or rolling circle PRINS, have been introduced [54, 55], which leads to think that the PRINS technology could provide an efficient complement to FISH and PCR in various diagnostic and research situations.

Acknowledgments

This study was supported by a European INTAS project (N° 03-51-4060).

References

[1] Pinkel, D., Straume, T., and Gray, J.M. (1986) Cytogenetic analysis using quantitative, high-sensitivity, fluorescence hybridization. *Proc. Natl. Acad. Sci. USA* 83, 2934-2938.

[2] Koch, J.E., Kolvraa, S., Petersen, K.B., Gregersen, N., and Bolund, L. (1989) Oligonucleotide-priming methods for the chromosome-specific labeling of alpha-satellite DNA in situ. *Chromosoma* 98, 259-265.

[3] Liehr, T., and Claussen, U. (2002) Current developments in human molecular cytogenetic techniques. *Curr. Molec. Med.* 2, 283-297.

[4] Levsky, J.M., and Singer, R.H. (2003) Fluorescence in situ hybridization: past, present and future. *J. Cell Sci.* 116, 2833-2838.

[5] Therkelsen, A.J., Nielsen, A., and Kolvraa, S. (1997) Localisation of the classical DNA satellites on human chromosomes as determined by primed in situ labelling (PRINS). *Hum. Genet.* 100, 322-326.

[6] Koch, J., Hindkjaer, J., Kolvraa, S., and Bolund, L. (1995) Construction of a panel of chromosome-specific oligonucleotide probes (PRINS-primers) useful for the identification of individual human chromosomes in situ. *Cytogenet. Cell Genet.* 71, 142-147.

[7] Pellestor, F., Girardet, A., Lefort, G., Andréo, B., and Charlieu, J.-P. (1995) Selection of chromosome specific primers and their use in simple and double PRINS technique for rapid in situ identification of human chromosomes. *Cytogenet. Cell Genet.* 70, 138-142.

[8] Jorgensen, A.L., Bostock, C.J., and Bak, A.L. (1987) Homologous subfamilies of human alphoid repetitive DNA on different nucleolus organizing chromosomes. *Proc. Natl. Acad. Sci. USA* 84, 1075-1079.

[9] Therkelsen, A.J., Nielsen, A., Koch, J., Hindkjaer, J., and Kolvraa, S. (1995) Staining of human telomeres with primed in situ labeling (PRINS). *Cytogenet. Cell Genet.* 68, 115-118.

[10] Krejci, K., and Koch, J. (1999) An in situ study of variant telomeric repeats in human chromosomes. *Genomics* 58, 202-206.

[11] Koch, J., Hindkjaer, J., Mogensen, J, Kolvraa, S., and Bolund, L. (1991) An improved method for chromosome-specific labeling of satellite DNA in situ by using denatured double-stranded DNA probes as primers in a primed in situ labeling (PRINS) procedure. *Genet. Anal. Tech. Appl.* 8, 171-178.

[12] Hindkjaer, J., Koch, J., Terkelsen, C., Brandt, C.A., Kolvraa, S., and Bolund, L. (1994) Fast, sensitive, multicolour detection of nucleic acids in situ by primed in situ labeling (PRINS). *Cytogenet. Cell Genet.* 66, 152-154.

[13] Gosden, J., and Lawson, D. (1995) Instant PRINS: a rapid method for chromosome identification by detecting repeated sequences in situ. *Cytogenet. Cell Genet.* 68, 57-60.

[14] Gosden, J., and Lawson, D. (1994) Rapid chromosome identification by oligonucleotide-primed in situ DNA synthesis (PRINS). *Hum. Mol. Genet.* 3, 931-936.

[15] Paskins, L., Brownie, J., and Bull, J. (1999) In situ polymerase chain reaction and cycling primed in situ amplification: improvements and adaptations. *Histochem. Cell Biol.* 111, 411-416.

[16] Harrer, T., Schwinger, E., and Mennicke, K. (2001) A new techniquefor cyclic in situ amplification and a case report about amplification od a single copy gene sequence in human metaphase chromosomes through PCR-PRINS. *Hum. Mutat.* 17, 131-140.

[17] Yan, J., Bronsard, M., and Drouin, R. (2001) Creating a new color by omission of 3 end blocking step for simultaneous detection of different chromosomes in multi-PRINS technique. *Chromosoma* 109, 565-570.

[18] Werner, M., Wilkens, L., Nasarek, A., Tchinda, J., and Komminoth, P. (1997) Detection of karyotype changes in interphase cells: oligonucleptide-primed in situ labelling versus fluorescence in situ hybridization. *Virchows Arch.* 430, 381-387.

[19] Gu, H.F., Lind, M.I., Wieslander, L., Landegren, U., Soderhall, K., and Melefors, O. (1997) Using PRINS for gene mapping in polytene chromosomes. *Chromosome Res.* 5, 463-465.

[20] Sola, L., Gabrielli, I., De Innocentiis, S., and Gornung, E. (1999) Chromosomal localization of zebrafish *Alu*I repeats by primed in situ (PRINS) labeling. *Cytogenet. Cell Genet.* 84, 28-30.

[21] Martins, C., and Galetti, P.M. (1999) Chromosomal localization of 5S rDNA genes in *Leporinus* fish (Anostomidae, Characiformes). *Chromosome Res.* 7, 363-367.

[22] Thomas, H.M., Williams, K., and Harper, J.A. (1996) Labelling telomeres of cereals, grasses and clover by primed in siyu DNA labelling. *Chromosome Res.* 4, 182-184.

[23] Abbo, S., Dunford, R.P., Miller, T.E., Reader, S.M., and King, I.P. (1993) Primer-mediated in situ detection of the B-hordein gene cluster on barley chromosome 1H. *Proc. Natl. Acad. Sci. USA* 90, 11821-11824.

[24] Pich, U., Meister, A., Macas, J., Dolezel, J., Lucretti, S., and Schubert, I. (1995) Primed in situ labelling facilitates flow sorting of similar sized chromosomes. *The Plant J.* 7, 1039-1044.

[25] Russo, A., Tommasi, A.M., and Renzi, L. (1996) Detection of minor and major satellite DNA in cytokinesis-blocked mouse splenocytes by a PRINS tandem labelling approach. *Mutagenesis* 11, 547-552.

[26] Gu, F., and Hindkjaer, J. (1996) Primed in situ labeling (PRINS) detection of the telomeric (CCCTAA)n sequences in chromosomes of domestic animals. *Mammal. Genome* 7, 231-232.

[27] Lavoie, J., Bronsard, M., Lebel, M., and Drouin, R. (2003) Mouse telomere analysis using an optimized primed in situ (PRINS) labeling technique. *Chromosoma* 111, 438-444.

[28] Speel, E.J.M., Lawson, D., Hopman, A.H.N., and Gosden, J. (1995) Multi-PRINS: multiple sequential oligonucleotide primed in situ DNA synthesis reaction label specific chromosomes and produce bands. *Hum. Genet.* 95, 29-33.

[29] Mennicke, K., Yang, J., Hinrichs, F., Muller, A., Diercks, P., and Schwinger, E. (2003) Validation of primed in situ labeling for interphase analysis of chromosomes 18, X and Y in uncultured amniocytes. *Fetal Diagn. Ther.* 18, 114-121.

[30] Pellestor, F., Girardet, A., Andréo, B., Lefort, G., and Charlieu, J.P. (1996) The PRINS technique: potential use for rapid preimplantation embryo chromosome screening. *Mol. Hum. Reprod.* 2, 135-138.

[31] Hindkjaer, J., Brandt, C.A., Stromkjaer, H., Koch, J., Kolvraa, S., and Bolund, L. (1996) Primed in situ labelling (PRINS) as a rational procedure for identification of marker chromosomes using a panel of primers differentially tagging the human chromosomes. *Clin. Genet.* 50, 437-441.

[32] Orsetti, B., Lefort, G., Boulot, P., Andréo, B., and Pellestor, F. (1998) Fetal cells in maternal blood: the use of primed in situ (PRINS) labelling technique for fetal cell detection and sex assessment. *Prenat. Diagn.* 18, 1014-1022.

[33] Krabchi, K., Gros-Louis, F., Yan, J., Bronsard, M., Masse, J., Forest, J.C., and Drouin, R. (2001) Quantification of all fetal nucleated cells in maternal blood between the 18th and the 22nd weeks of pregnancy using molecular cytogenetic techniques. *Clin. Genet.* 60, 145-150.

[34] Coullin, P., Andréo, B., Charlieu, J.P., Candelier, J.J., and Pellestor, F. (1997) Primed in situ (PRINS) labelling with Alu and satellite primers for rapid characterization of human chromosomes in hybrid cell lines. *Chromosome Res.* 5, 307-312.

[35] Callen, D.F., Yip, M.Y., and Eyre, H.J. (1997) Rapid detection of euchromatin by Alu-PRINS: use in clinical cytogenetics. *Chromosome Res.* 5, 81-85.

[36] Wilkens, L., Komminoth, P., Nasarek, A., Von Wasielewski, R., and Werner, M. (1997) Rapid detection of karyotype changes in interphase bone marrow cells by oligonucleotide primed in situ hybridization(PRINS). *J. Pathol.* 181, 368-373.

[37] Ramael, M., Van Steelandt, H., Styven, G., Van Steenkiste, M., and Degroote, J. (1997) Application of the primed in situ labeled (PRINS) method for detection of numerical chromosomal aberration in paraffin embedded formalin fixed tissue of molar and non-molar pregnancies. *Biochemica* 2, 18-20.

[38] Mogensen, J., Kolvraa, S., Hindkjaer, J., Petersen, S., Koch, J., Nygard, M., Jensen, T., Gregersen, N., Junker, S., and Bolund, L. (1991) Nonradioactive, sequence-specific detection of RNA in situ by primed in situ labeling (PRINS). *Exp. Cell Res.* 196, 92-98.

[39] Bains, M.A., Agarwal, R., Pringle, J.H., Hutchinson, M.R., and Lauder, I. (1993) Flow cytometric quantitation of sequence-specific mRNA in hemopoietic cell suspensions by primer-induced in situ (PRINS) fluorescent nucleotide labeling. *Exp. Cell Res.* 208, 321-326.

[40] Andersen, C.L., Koch, J., and Kjeldsen, E. (1998) CpG islands detected by self-primed in situ labeling (SPRINS). *Chromosoma* 107, 260-266.

[41] Krejci, K., and Koch, J. (1998) Improved detection and comparative sizing of human chromosomal telomeres in situ. *Chromosoma* 107, 198-203.

[42] Krejci, K., and Koch, J. (1999) An in situ study of variant telomeric repeats in human chromosomes. *Genomics* 58, 202-206.

[43] Coullin P., Philippe, C., Ravise, N., and Bernheim, A. (1999) Simultaneous fluorescence in situ hybridization (FISH) and R-banding by primed in situ labelling (PRINS). *Chromosome Res* 7, 241-242.

[44] Pellestor, F., Quenesson, I., Coignet, L., Girardet, A., Andréo, B., Lefort, G., and Charlieu, J.P. (1996) FISH and PRINS, a strategy for rapid chromosome screening: application to the assessment of aneuploidy in human sperm. *Cytogenet. Cell Genet* 72, 115-118.

[45] Pellestor, F., Girardet, A., Coignet, L., Andréo, B., and Charlieu, J.P. (1996) Assessment of aneuploidy for chromosomes 8, 9, 13, 16 and 21 in human sperm by using primed in situ labeling technique. *Am. J. Hum. Genet.* 58, 797-802.

[46] Pellestor, F., Girardet, A., Coignet, L., Andréo, B., Lefort, G., and Charlieu, J.P. (1997) Cytogenetic analysis of meiotic segregation in sperm from two males heterozygous for

reciprocal translocations using PRINS and humster techniques. *Cytogenet. Cell Genet.* 78, 202-208.

[47] Pellestor, F., Malki, S., Andréo, B., and Lefort, G. (2002) Ultra-rapid multicolor PRINS protocol for chromosome detection in human sperm. *Chromosome Res* 10, 359-367.

[48] Pellestor, F., Anahory, T., Andréo, B., Régnier-Vigouroux, G., Soulié, J.P., Baudouin, M., and Demaille, J. (2004) Fast multicolour primed in situ protocol for chromosome identification in isolated cells may be used for human oocytes and polar bodies. *Fertil. Steril.* 81, 408-415.

[49] Troyet, D.L., Goad, D.W., Xie, H., Rohrer, G.A., Alexander, L.J., and Beattie, C.W. (1994) Use of direct in situ single copy (DISC) PCR to physically map 5 porcine microsatellites. *Cytogenet. Cell Genet.* 67, 199-204.

[50] Kadandale, J.S., Wachtel, S.S., Tunca, Y., Wilroy, R.S., Martens, P.R., and Tharapel, A.T. (2000) Localization of SRY by primed in situ labeling in XX and XY sex reversal. *Am. J. Hum. Genet.* 95, 71-74.

[51] Cinti, C., Stuppia, L., and Maraldi, N.M. (2002) Combined use of PRINS and FISH in the study of the dystrophin gene. *Am. J. Med. Genet.* 107, 115-118.

[52] Tharapel, A.T., Kadandale, J.S., Martens, P.R., Wachtel, S.S., and Wilroy, R.S. (2002) Prader Willi/Angelman and DiGeorge/Velocardiofacial syndrome deletions: diagnosis by primed in situ labeling (PRINS). *Am. J. Med. Genet.* 107, 119-122.

[53] Tharapel, S.A.,and Kadandale, J.S. (2002) Primed in situ labeling (PRINS) for evaluation of gene deletions in cancer. *Am. J. Med. Genet.* 107, 123-126.

[54] Nilsson, M., Krejci, K., Koch, J., Kwiatkowski, M., Gustavsson, P., and Landegren, U. (1997) Padlock probes reveal single-nucleotide differences, parent of origin and in situ distribution of centromeric sequences in human chromosomes 13 and 21. *Nat. Genet.* 16, 252-255.

[55] Larsson, C., Koch, J., Nygren, A., Janssen, G., Raap, A.K., Landegren, U., and Nilsson, M. (2004) In situ genotyping individual DNA molecules by target-primed rolling-circle amplification of padlock probes. *Nat. Methods* 1, 227-232.

In: PRINS and PNA Technologies in Chromosomal Investigations ISBN 1-60021-413-4
Editor: Franck Pellestor, pp. 13-21

Chapter 2

Simultaneous Detection of DNA and Protein Targets by PRINS Combined with Immunohistology

Philippe Coullin*
Endocrinologie et génétique de la reproduction et du dévelopement,
Clamart, France

Abstract

Primed *in situ* labeling (PRINS) has proven to be an attractive alternative to fluorescence *in situ* hybridization (FISH) for *in situ* DNA labeling and for being combined on the same slide with others methods. The simultaneous visualization of a DNA and a protein target is an important goal useful for example in viral detection, tagging cells of interest and chromosome physiology investigations. For reasons of rapidity, probe accessibility and protein structure preservation, PRINS may be an extra advantage in such combined procedures. Such successful realizations for several items, following adapted protocols, show the great adaptive plasticity of the PRINS-immunostaining combination useful in cytogenetic preparations or in tissue sections.

Key Words: PRINS, immunostaining

Introduction

Koch et al. [1] and Gosden et al. [2] proposed a method for chromosome labeling based on rapid annealing of unlabelled DNA primers to complementary target sequences, followed by *in situ* primer elongation with Taq polymerase. The method called primed *in situ* labeling

* Correspondence to: Dr Philippe Coullin, INSERM U782, Endocrinologie et génétique de la reproduction et du dévelopement, 32 rue des carnets, 92140 Clamart, France; E-mail: philippe.coullin@inserm.ipsc.u-psud.fr

(PRINS) has become in the past decade an alternative to fluorescent *in situ* hybridization (FISH). It was proved that PRINS could work on single copy gene by specific adaptations like using several primers and tyramide signal amplification [3, 4] However, its best efficiency is on targets containing repetitive sequences. Despite this limitation, PRINS proved useful because it generates highly specific labeling and it can be performed rapidly and easily with cost-effectiveness. Also, PRINS can be used on tissue sections and it can be combined with other methods like fluorescent *in situ* hybridization (FISH) and immunostaining [5].

From FISH to PRINS

The simultaneous visualization of a DNA and a protein target is an important goal. In a series of reports, as for example by Funabiki et al. [6] studying the cell cycle dependent specific positioning and clustering of centromeres and telomere in fission yeast, FISH has been combined with immunocytochemistry. However, only a few reports associated FISH and immunofluorescence. Developing such methodology, Saffery et al. [7] have extensively characterized the proteins associating with two different homologue human neocentromeres in interphase and prometaphase/metaphase, and compared these directly with those found with normal human centromeres. They observed that human neo and centromeres show identical distribution patterns of > 20 functionally important kinetochore-associated proteins. In another domain, Deb et al. [8] showed that bone marrow-derived myofibroblasts are present in adult human heart valves. On heart valves of gender-mismatched bone marrow transplant patients, they investigated the presence of chimeric cells expressing calponin, a smooth muscle-specific protein, and the patient/donor origin of cells by tagging for X and Y chromosomes by FISH.

From these examples, it could be seen the interest to efficiency combine protein and nucleic acid detection in the same biological material as well in cytogenetic preparations and tissue sections. Protocols have been developed for various purposes: study the structural organization of the chromosome and the nucleus, determine the cytokinetic parameters of tumor cell populations, detect the immunophenotype of cells harboring a specific chromosomal aberration or virus infection, tag the origins of cells of interest. Both the success and the sensitivity of such combined procedures depend on factors like the preservation of the cell morphology and protein epitopes, the accessibility of the targets, the lack of cross-reactions, the stability of the staining [9, 10]. However, for reasons of rapidity, probe accessibility and to avoid formamide for a long hybridization time, the substitution of FISH by PRINS may be an extra advantage in such combined procedures.

PRINS had been successively combined with immunochemistry for investigations in different domains:

a) The structure and the physiology of chromosomes. Thus, Mitchell et al. (11) investigated the organization of repetitive DNA sequences with respect to kinetochores. They clearly showed that alphoid DNA sequences are closely associated with kinetochores of human chromosomes. Simple-sequence satellite

DNAs occupies separate, non-overlapping domains within the centromere. These two major families are separated by a third, relatively low-copy repetitive DNA family.

b) Viral detection as performed by Ramael et al. (12), for SV40 like viral DNA and antigens in malignant pleural mesothelioma.
c) Tagging cells of interest (mainly in xenotransplantion investigations) as did Martin et al. (13) for determining the origin of the cells. They used PRINS to confirm the presence of bovine cells within the porcine skeletal muscle tissue section in their transplation experiments, showing that bone marrow stromal cells can participate in the development of skeletal muscle.

Methods

In fact, there are as numerous techniques as data reported. Three technical protocols are extensively described in [14] and [15]. Each procedure has been adapted to the item purchased with some variations: PRINS before or after immunological process, fluorescence or immunochemistry, cytogenetic like slides or tissue sections.

The first step arise the problem of fixation of the sample preserving both antigenic specificity of the proteins and DNA integrity. As PRINS is working by elongation using Taq polymerase, it is important that the DNA was not broken. It is to notice that probably for this raison, in classical cytogenetic preparation, the elongation provided by PRINS is limited to about 100pb. For cytogenetic preparations, simple air-drying associated with either a cold methanol, ethanol, acetone solutions or a classical but quick methanol-acetic acid were proposed with success. Frozen tissue sections with this quick cytogenetic fixative are compatibles with PRINS. PRINS could also work on paraffin embedded section. In this case, the fixative protocol seems of importance and a 3% formaldehyde solution well buffered at pH 7.3 give good results.

Another problem may be the preservation of antigenic epitopes during the high temperature steps required by PRINS for *in situ* DNA denaturation and elongation. In most cases, the investigators choose to do the immunological reaction first and the PRINS in second with sometime a step of fixation between reactions which may be drying or using a formaldehyde solution. In this last case a permeabilisation is performed, using for instance Proteinase K. Other experimented a cold denaturation in NaOH solutions. In our own experiences, we performed PRINS first and indirect immunofluorescence (IIF) in second, with also very good results. This diversity demonstrates the great adaptive plasticity of the PRINS-immunostaining combination.

Assays on Chromosome Physiology

We present here our experiences combining PRINS and indirect immunofluorescence (IIF), developed for the study of chromosome segregation during abnormal mitoses, particularly during the megakaryocyte maturation to platelets [15, 16].

In contrast to malignant cells, few normal mammalian cells have a polyploid DNA content. Bone marrow megakaryocytes, the platelet precursors, are polyploid and this DNA increase is reached through an original process called endomitosis, which occurs during terminal differentiation. The main consequence of megakaryocyte polyploidisation is to augment cell size and consequently to increase the production of platelets which originate from the megakaryocyte cytoplasmic fragmentation. It was also hypothesized that the level of megakaryocyte ploidy modifies platelet size and functions by amplifying the genome and altering gene regulation of this highly specialized cells.

Elucidation of the endomitosis mechanisms and their regulation appears essential for improving our knowledge of megakaryocyte biology and for understanding the advantages of such a division pattern. Megakaryocytes undergo a normal cycle progression with G1, S, and G2 phases; the first stages of mitosis are preserved as well. The first stage of anaphase normally occurs. In contrast, the second part corresponding to spindle-pole to pole elongation and chromosome segregation via microtubule retraction seems absent or very limited. Thus, megakaryocyte endomitosis are interrupted in anaphase.

One critical event during normal mitotic process is an accurate chromosome segregation, which ensures an equivalent DNA repartition between the two daughter cells. This crucial process, is monitored by a powerful mitotic checkpoint that is able to delay the metaphase to anaphase transition in case of case of spindle assembly abnormality or a failure in proper chromosome attachment. However, by centromeric PRINS (or FISH), it was easily observed that the different chromosomes were not equally distributed within the different lobe of the megakaryocytic nucleus (Fig. 1). We investigated how chromosome segregation occurred on a multipolar spindle in endomitotic megakaryocytes. This is why we developed PRINS-IIF methodology in order to simultaneously visualize by fluorescence both the spindle and selected centromeres.

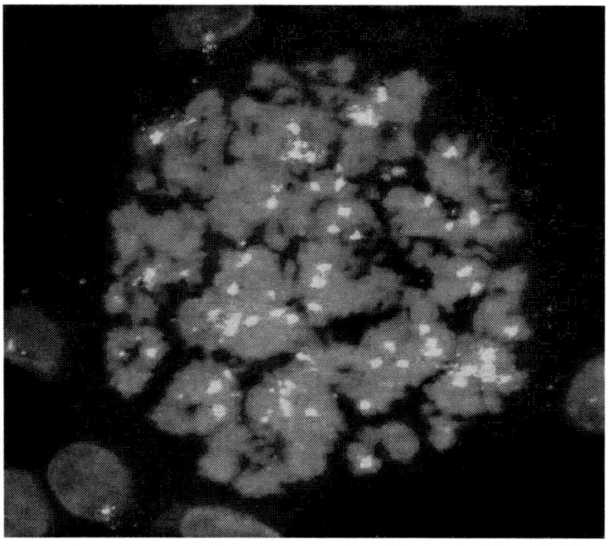

Figure 1. Asymmetrical distribution of the chromosomes within the different lobe of the megakaryocytic nucleus of high ploidy. Demonstration by PRINS (Fluorecein-12-dUTP: green; DAPI DNA counterstain: blue) using here primers specific for chromosome 1

Surprisingly our findings demonstrated that the mitotic checkpoint was completely functional and that the asymmetrical repartition of the chromosomes might be a consequence of the spindle organization and not sister –chromatids separation. After some wastage with FISH, we speculated that PRINS could provide a more advantageous approach and consequently, we used the methodology previously described [15] and illustrated in figure 2. We tested different possibilities and found that the tubulin epitope was far better preserved when immunostaining was performed after DNA labeling. In order to stabilize the immunological signal, Speel et al. [17] who performed PRINS in a second step, used a fluorescent alkaline phosphatase (Apase)-Fast Red immunostaining method. They so enabled the simultaneous detection of diverse kind of surface antigens and repeated chromosome-specific DNA in somatic cell hybrid and tumor cell lines.

For PRINS labeling, we observed that direct fluorescent incorporation gave very good staining, avoiding the incubation and washings required by an indirect procedure and saving time. The dehydration step between PRINS and immunostaining reactions seems of importance. However a high temperature (about 70°C) is two times required: one time for the denaturation of the chromosomes (it should be idem for FISH) and another time for the elongation. This implies that the epitope recognized by the anti-tubulin must be heat-resistant. Fortunately, it appeared to be the case. The denaturation procedure was classically conducted in a 70 % formamide /30% 2X solution, pH 7-7.2 at 71°C for 2 min and the elongation time was 15 min at 72 °C. In order to avoid the high denaturation temperature, Mitchell [18] also experimented a cold denaturation in a NaOH solution.

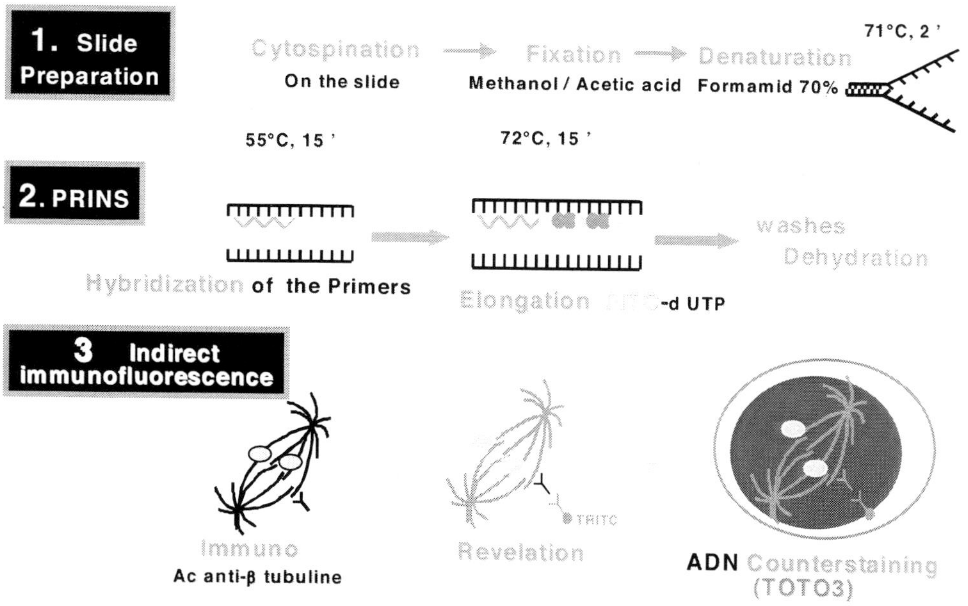

Figure 2. Schematized technical PRINS-IIF procedure

In our preparations, interphasic nuclei contained green (FITC) dots, corresponding to centromeres of the selected chromosomes. In this study, we used primers respectively specific for the centomeres of human chromosomes 1 and 7 as previously described [19].

Two copies were observed when the cells were diploid and 2^XN copies when their DNA content was polyploid. During mitosis, these dots were dispersed on the spindle labeled in red (TRITC) by immunofluorescence. A classical IIF procedure was conducted using an anti-β-tubulin antibody (monoclonal; mouse) as first antibody and a donkey TRITC-labeled anti-mouse F(ab')2 as second antibody. The whole DNA was counterstained in blue using TOTO 3 and DAPI (Fig. 3). Four dots for diploid cells were present when sister-chromatids separated from the beginning of anaphase, to one each side of the equator. In case of polyploid cells, we observed complex spherical multipolar spindles. The total number of labeled centromere was always even as would be expected from synchronous and perfectly regulated chromosome segregation. However, we immediately observed, that these labeled centromeres, were not symmetrically distributed along the different components of the multipolar spindle in all endomitotic cells examined. Some spindle-poles did not bear any of the selected-labeled chromosomes (Fig. 4).

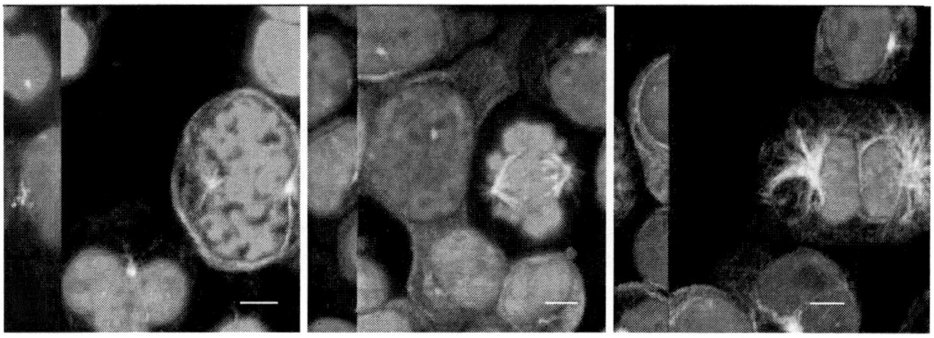

Figure 3. PRINS and IIF procedure on interphase and different normal metaphase steps. PRINS (Fluorecein-12-dUTP: green) using primers specific of chromosome 7. IIF, specific of the anti- -tubulin antibody as first antibody and a donkey TRITC-labeled anti-mouse, evidencing the spindle in red. Whole DNA counterstain by DAPI: blue

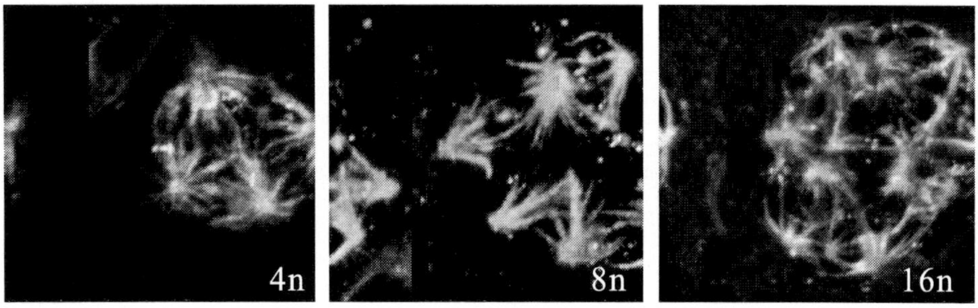

Figure 4. Asymmetrical distribution of the chromosomes within the multipolar spindle of polyploid megacaryocytes suggesting an asymmetrical repartition at anaphase steps. PRINS (Fluorecein-12-dUTP: green) using primers specific of chromosome 1. IIF, specific of the anti- β -tubulin antibody as first antibody and a donkey TRITC-labeled anti-mouse, evidencing the spindle in red. The acquisitions were realised using a confocal microscope

These observations demonstrate that chromosome repartition is performed asymmetrically on the multipolar spindle of megakaryocytes, regardless of the chromosome category. The total number of chromosome at the cell level is correct, but the repartition in each nuclear lobe is aleatoric. For a technical point of view, our labeling protocol gave very reproducible results, demonstrating the feasibility of both direct fluorescent PRINS and immunofluorescence and the efficiency of PRINS for studying chromosome segregation.

Tagging the Origin of Cells of Interest on Tissue Sections

In the study of Religa et al. [20], the problem was to determine if smooth-muscle progenitor cells of bone marrow origin contribute to the development of neo-intimal thickenings in rat aortic allografts and injured carotid arteries. To explore this possible bone marrow origin, female LEW rats were irradiated and substituted with bone marrow from male LEW rats. Subsequently, the animals received an aortic transplant from female F344 rats or were exposed to balloon injury of the carotid artery. Combined PRINS labeling and immunostaining contributed to verify that not only inflammatory but also smooth muscle-like cells of male origin appeared in the vessel wall in both situations. These observations suggested that the smooth-muscle cells participating in the development of neo-intimal lesions during vascular disease might, in part, originate from circulating progenitors.

These simultaneous stainings were conducted on formaldehyde fixed and embedded paraffin tissue sections, with specific permeabilisation procedures before the two steps. Conversely as that we did, the immunological process was realized first, using SM α-actin, SM myosin or leukocyte (CD45) antibodies. In order to preserve the staining, second formaldehyde fixation was done before performing PRINS. This, by the way, illustrates the feasibility of PRINS not only on frozen but also in fixed tissue sections. The cells were then tagged for the presence of *SRY* gene using a couple of primer and a biotinylated dCTP. It is to notice that the PRINS procedure was repeated five times as a cycling PRINS. Detection of amplified signal signals from DNA and proteins were performed simultaneously using Cy2 steptavidin and Cy5 anti-mouse antibody. The protocol developed in this study also indicated that PRINS was efficient for the detection of non-highly repetitive targets.

Conclusion

The combination of FISH/PRINS and immunostaining offers multiple possibilities for research mainly in viral detection, tagging cells of interest and chromosome physiology. Yet the simultaneous generation of signals involving DNA and protein staining may be difficult. Among the problems encountered, there are a) the persistence of immunological signals after FISH/PRINS treatments, b) the permeability of DNA probes after protein fixation, and c) the antigen durability after DNA denaturation. However, FISH/PRINS and protein immunostaining have been combined in a significant number of studies following various protocols. In our study of chromosome distribution and segregation during the polyploidisation process of megakaryocytes, we developed a simple method combining PRINS with labeling of selected centromeres and fluorescent immunostaining of tubulin in order to visualize the spindle. From our experience and other reports, it could be speculated that PRINS would prove more advantageous than FISH for several reasons: a) PRINS reactions are highly specific b) the PRINS procedure is rapid c) the prolonged contacts with formamide required during FISH incubation are avoided in PRINS protocols d) the diffusion of small oligonucleotides within the material is more efficient than the *in situ* diffusion of FISH probes.

In conclusion, PRINS proved to be an attractive technology and an excellent alternative to FISH for achieving fluorescent *in situ* DNA labeling combined with the indirect immonufluorescent/immunochemistry tagging of proteins in cytogenetic like preparations or in tissue sections.

References

[1] Koch, J., Kolvraa, S., Gregersen, N., Petersen, K.B. and Bolund, L. (1988). PRINS and PAL-two new techniques for the visualisation of specific DNA sequences. *Genome* 30, suppl.1: 446.

[2] Gosden, J., Hanratty, D., Starling, J., Fantes, J., Mitchell, A. and Porteous, D. (1991). Oligonucleotide-primed in situ DNA synthesis (PRINS): a method for chromosome mapping, banding, and investigation of sequence organization. *Cytogenet Cell Genet* 57: 100-104.

[3] Kadandale, J.S., Wachtel, S.S., Tunca, Y., Martens, P.R., Sid Wilroy, R. and Tharapel, A.T. (2002). Deletion of RBM and DAZ in azoospermia: evaluation by PRINS. *Am j Med genet* 107: 105-108.

[4] Krabchi, K., Lavoie, J., Coullin, P., Bronsard, M., Pellestor, F., Yan, J. and Drouin, R. (2004). De la conception du PRINS à son couronnement. [From the conception of the PRINS to its coronation]. *Med Sci (Paris*). 20:465-473.

[5] Coullin, P., Roy, L., Pellestor, F., Candelier, J.J., Bed'Hom, B., Guillier-Gencik, Z. and Bernheim, A. (2002). PRINS, the other *in situ* DNA labeling useful for cellular biology. *American journal of medical genetics* 107:127-135.

[6] Funabiki, H., Hagan, I., Uzawa, S. and Yanagida, M. (1993). Cell cycle-dependent specific positioning and clustering of centromeres and telomeres in fission yeast. *J Cell Biol.* 121:961-976.

[7] Saffery, R., Irvine, D.V., Griffiths, B., Kalitsis, P., Wordeman, L. and Choo, K.H. (2000). Human centromeres and neocentromeres show identical distribution patterns of >20 functionally important kinetichore-associated proteins. *Human Molec Genet* 9: 175-185.

[8] Deb. A., Wang, S.H., Skelding, K., Miller, D., Simper, D. and Caplice, N. (2005). Bone marrow-derived myofibroblasts are present in adult human heart valve. *J Heart Valve Dis.* 14(5):674-678.

[9] Speel, E.J.M., Herbergs, J., Ramaekers, F.C.S. and Hopman, A.H.N. (1994) Combined immunocytochemistry and fluorescence *in situ* hybridization for simultaneous tricolor detection of cell cycle, genomic, and phenoypic parameters of tumor cells. *J.histochem.cytochem* 42: 961-966.

[10] Speel, E.J.M. (1999). Detection and amplification systems for sensitive, multiple-target DNA and RNA *in situ* hybridization: looking inside cells with a spectrum of colors. *Histohem.Cell Biol* 112: 89-113.

[11] Mitchell, A., Jeppesen, P., Hanratty, D. and Gosden, J. (1992). The organisation of repetitive DNA sequences on human chromosomes with respect to the kinetochore analysed using a combination of oligonucleotide primers and CREST anticentromere serum. *Chromosoma.* 101:333-341.

[12] Ramael, M., Nagels, J., Heylen, H., de Schepper, S., Paulussen, J., de Maeyer, M. and van Haesendonck, C. (1999). Detection of SV40 like viral DNA and viral antigens in malignant pleural mesothelioma. *Eur Respir J.* 14:1381-1386.

[13] Martin, P.L., Davis, D., Weiss, M., Grieger, D., Abou-Easa, K. and Troyer, D. (2001). Xenotransplantation of bovine bone marrow stromal cells into pig fetuses: incorporation into skeletal muscle. *Anim Biotechnol* 12:183-191.

[14] Speel, E.J.M., Ramaekers, F.C.S. and Hopman, A.H.N. (2003). Primed *in situ* nucleic acid labeling combined with immunocytochemistry to simultaneously localize DNA and protein in cells and chromosomes. *Methods in molecular biology.* 226: 453-464. PCR Protocols, Second Edition. Edited by JMS Bartlett and D Stirling. Humana Press inc., Totowa, NJ, USA.

[15] Coullin, P. (2006). PRINS combined with indirect immunofluorescence. *Methods in molecular Biology. Book Series Format Guide.* « PRINS and *in situ* PCR Protocols », (second edition). Cha 3:23-32. Edited by Franck Pellestor. The Human Press Inc., Totowa, NJ, USA.

[16] Roy. L., Coullin, P., Vitrat, N., Hellio, R., Debili, N., Weinstein, J., Bernheim, A. and Vainchenker, W. (2001). Asymetrical segregation of chromosomes with a normal metaphase/anaphase checkpoint in polyploid megakaryocytes. *Blood* 97:2238-2247.

[17] Speel, E.J.M., Lawson, D., Ramaekers, F.C.S., Gosden, J.R. and Hopman, A.H.N. (1997). Combined immunochemistry and PRINS DNA synthesis for simultaneous detection of phenotypic and genomic parameters in cells. *Methods in molecular Biology.* « PRINS and *in situ* PCR Protocols », 71:53-60. Edited by JR Gosden. The Human Press Inc., Totowa, NJ, USA.

[18] Mitchell, A.R. (1997). Chromosomal PRINS DNA labelling combined with indirect immunocytochemistry. *Methods in molecular Biology.* « PRINS and *in situ* PCR Protocols », 71:61-70. Edited by JR Gosden. The Human Press Inc., Totowa, NJ, USA.

[19] Pellestor, F., Girardet, A., Lefort, G., Andréo, B. and Charlieu JP. (1995). Selection of chromosome-specific primers and their use in simple and double PRINS techniques for rapid in situ identification of human chromosomes. *Cytogenet Cell Genet* 70: 138-142.

[20] Religa, P., Bojakowski, K., Maksymowicz, M., Bojakowska, M., Sirsjo, A., Gaciong, Z., Olszewski, W., Hedin, U. and Thyberg J. (2002). Smooth-muscle progenitor cells of bone marrow origin contribute to the development of neointimal thickenings in rat aortic allografts and injured rat carotid arteries. *Transplantation* 9:1310-1315.

In: PRINS and PNA Technologies in Chromosomal Investigations ISBN 1-60021-413-4
Editor: Franck Pellestor, pp. 23-35

Chapter 3

PRINS for Detection of Unique Chromosome Sequences and Single Gene Deletions

Stephen S. Wachtel* and Avirachan T. Tharapel
University of Tennessee, Memphis
Memphis, Tennessee, 38163, U.S.A.

Abstract

By *PRINS,* unlabeled oligonucleotide primers are annealed to denatured complementary sequences on chromosomes *in situ,* and extended in the presence of labeled nucleotides and Taq DNA polymerase. Chromosomal DNA acts as template for the extension, and the labeled nucleotides act as substrate for the DNA polymerase. The labeled extended target is visualized by fluorescence microscopy. Thus PRINS can be used to detect subtle chromosome rearrangements and microdeletions that are not amenable to analysis by standard analytical methods such as FISH. The PRINS method can also be used to identify single copy sequences and individual genes.

Keywords: Primed in situ labeling, PRINS, chromosome, gene mapping, single-copy gene, microdeletion

Introduction

Primed in situ labeling (PRINS) combines aspects of the polymerase chain reaction (PCR) and fluorescence in situ hybridization (FISH) for the localization of repetitive and unique nucleotide sequences on chromosomes. The PRINS signals, generated by annealing

* Correspondence to: Prof. Stephen Wachtel, Department of Obstetrics and Gynecology, University of Tennessee 853 Jefferson Ave. Memphis TN 38163 USA. E-mail: swachtel@utmem.edu

and subsequent extension of oligonucleotide primers, are identified and scored by fluorescence microscopy. Although the method was described in 1989 (Koch et al. 1989) it has not been used widely, ostensibly because of technical difficulties experienced in its application. Yet PRINS offers a unique, rapid approach to labeling, and could be used to identify any DNA sequence for which one or more oligonucleotide primers are available—i.e. any published sequence.

Anomalies of chromosome structure and number generate a variety of abnormal phenotypes including physical malformations, mental retardation and infertility. This is significant because some 0.7% of live-born children exhibit congenital malformations secondary to visible chromosome anomalies (Nussbaum et al., 2004). For that reason, fine point analysis of the chromosomes is a significant factor in the study of congenital diseases, with implications for clinical management and counseling. FISH has been an important and useful adjunct to the cytogenetic evaluation of chromosome rearrangements and gene deletions and duplications for more than a quarter century. FISH enables resolution of DNA sequences separated by only a few megabases in metaphase or prophase chromosomes (the low-end resolution is approximately 20 kb). Yet production of locus-specific FISH probes can be tedious and costly, and many of the probes are not commercially available. By contrast, PRINS is fast and relatively inexpensive, and the probes can be synthesized routinely.

Although most early applications of the PRINS method dealt with repetitive sequences such as the alpha-satellite, *Alu* and telomeric repeats (Pellestor et al., 1995; 1994; Gosden et al., 1991), PRINS can be used to identify single copy genes. In fact PRINS was used to identify the X-linked *Factor IX* gene more than 10 years ago (Cinti et al., 1993). That study was the first published example of identification of a single copy gene by PRINS, and it represents a significant biologic milestone, because every known gene sequence is a potential primer source. It follows that any gene or exon could be located in chromosome preparations *in situ,* provided that a unique nucleotide sequence is available.

In this chapter, we describe our experience with the PRINS method, with special emphasis on the detection of (i) unique sequences and single copy genes such as *SRY* and *SOX3* and (ii) microdeletions involving the *p53, RB1, SNRPN, GABRB3* and *DGCR2/TUPLE1* loci.

The Basic Method

In brief, the PRINS technique comprises four steps: (i) denaturation of chromosomal DNA, (ii) hybridization of primers *(annealing),* (iii) extension of DNA chains in the presence of labeled nucleotides and Taq DNA polymerase, and (iv) detection of the newly-synthesized labeled DNA by use of antibodies attached to fluorescent dyes. In our experience, a typical PRINS reaction mixture has a volume of 50 μl and contains: 200-500 pmol of primer, 0.2 mM dATP, dCTP and dGTP, 0.02 mM dTTP, 0.02 mM bio-16-dUTP, digoxygenin-11-dUTP, or fluorescein dUTP, 50 mM Tris-HCl (pH 8.3), 1.5 mM $MgCl_2$, 50 mM KCl, 0.01% bovine serum albumin and 2 units Taq polymerase. The reaction mixture is placed on the denatured chromosome preparation. Denaturation is accomplished by immersing slides in

70% formamide-2 x SSC for 2-3 minutes at 72°C, and dehydrating them through a series of cold ethanol washes. After the slides are air-dried, the reaction mixtures are layered onto the slides, and overlaid with coverslips that are sealed in place with rubber cement. The slides are put on a flat plate programmable thermal cycler and maintained at annealing temperatures for 5-10 minutes followed by extension for 20-30 minutes at 70-72°C. The reaction is stopped by immersing the slides in 500 mM NaCl-50 mM EDTA, pH8 for 5 minutes at 72°C. Labeled sequences are detected by immunocytochemistry and fluorescence microscopy. Biotin-complexed fragments can be visualized by use of fluoresceinated avidin, and digoxigenin, by use of fluoresceinated digoxigenin antibody. The chromosome preparations are counterstained with propidium iodide or DAPI, and scored under ultraviolet light.

Although our reactions are carried out on a programmable thermal cycler equipped with a flat plate for slides (HYBrite, Vysis Inc., Downers Grove IL), annealing and extension can readily be accomplished on thermoblocks or even in metal containers suspended in hot water baths (Koch et al. 1995). Of course temperature is a critical factor under these conditions, and must be carefully controlled. For detection of single copy genes and unique sequences, we used a modified PRINS method incorporating (i) multiple primers for the same locus, (ii) single-step annealing and extension, (iii) anti-Taq DNA polymerase antibody and (iv) stringent washing. A comprehensive description of the modified PRINS method is given in Tharapel and Wachtel (in press).

PRINS Primers

Oligonucleotide primers were synthesized and HPLC-purified at Research Genetics, Huntsville AL. The primers were stored at –20°C.

After counting the individual A, C, G and T nucleotide residues of the primers, annealing temperatures may be computed by use of either of the following formulas:

$$T_M = 69.3 + 0.41\ (\%\ G + C) - 650\ /\ L$$

where L (length) = the total number of nucleotides in the primer.

$$T_M = 4\ (G + C) + 2\ (A + T)$$

When different temperatures are obtained, the results may be averaged. In general, satisfactory annealing occurs at temperatures between 55°C and 72°C. The higher temperatures increase annealing specificity.

The sequences of the PRINS primers are described in the various citations and in Table 1. When available, standard FISH probes were used to validate the data obtained with the PRINS primers.

Table 1. Primer sequences from cited studies

Locus	Sequence 5'—3'	Reference
RBM	GACCACCAGCTTCTTCGAGA GGATGAACATTCCTCTAGAGG GCCCTTCAGGAAGTCTGAGA AGAGGCCATGCATACCGTGA	Kadandale et al. (2002)
DAZ	CTCTGCCTCTGGCTTTACCA GAGGAGGCATCTGGAAATCATT GGAAGCTGCTTTGGTAGATAC TAGGTTTCAGTGTTTGGATTCCG	Kadandale et al. (2002)
SNRP	ACGTACCAGAGGTGGAAGTC GTCTTCAGAAGCATCAAGTTT ATGACTGTGGAGGGGCCAC AGGCATTATGGCTCCTCCAC	Tharapel et al. (2002)
GABRB3	TTCTGCAGGCACCGTCGGG AGAGGTCCGAGTCCCGTG AGGAAGGCTTTTCGGCATCT TAAGTGGGCTGCGCGCGGC	Tharapel et al. (2002)
DGCR/TUPLE1	GCGGTGCAACCCTGGGCA CAGTATGTTATCACTGGCCGG CGCCTGGTCGTCAGCTGC CAACCACAATGGCAAGCCGA AAGTTCCCAGAAATTCTAGCTA GTCATCCATGAACTGTTTGAC	Tharapel et al. (2002)

PRINS in Detection of the *SRY* Gene

The *SRY (sex-determining region Y)* gene, discovered by Sinclair et al. (1990), is the switch that determines the sex of the mammalian embryo. In the presence of *SRY,* the indifferent embryonic gonad becomes a testis. The testis secretes testosterone and antimullerian hormone and the male body type is induced. In the absence of *SRY,* the gonad becomes an ovary. In the absence of testosterone and antimullerian hormone, the female body type is induced.

Errors in the expression of *SRY* accordingly generate errors in the development of the sex phenotype in humans and in mammals generally. For example translocation of *SRY* from the Y to the X chromosome in the male germ line can cause development of the male phenotype in the absence of the Y chromosome *(XX male syndrome).* And deletion or mutation of *SRY* can cause development of the female phenotype in the presence of the Y chromosome *(XY gonadal dysgenesis)* (XY gonadal dysgenesis can also occur in the presence of *SRY* and XX male syndrome can occur in the absence of *SRY)* (Wachtel, in press; and see below).

Study of the *SRY* gene can accordingly provide insights into the etiology and nature of various sex-reversed syndromes with implications for theory and clinical management. The *SRY* gene is located on the short arm of the human Y chromosome, in distal Yp11.2, close to the pseudoautosomal region. It is 2.1 kb in length and contains a single exon, including a conservative motif found among the high mobility group (HMG) proteins. The conservative motif spans 79 amino acids that are associated with DNA binding. This supports the presumptive role of the SRY protein as a transcription factor.

Using the PRINS method, we studied the *SRY* gene in two 46,XX males, in a woman with 46,XY gonadal dysgenesis, and in an azoospermic man with an Xp-Yp interchange (Kadandale et al., 2000). One of the XX males was a 30-year-old who had been referred for small testes and infertility—hallmarks of the XX male syndrome. The other XX male was a newborn referred for small phallus, third degree hypospadias, and bifid scrotum with labioscrotal folds. The scrotum contained abnormally small testes. These abnormalities are encountered in some 30% of XX males. The XY female was evaluated originally for primary amenorrhea. She was 19 years old, and 1.65 m tall with normal breast development but her uterus was hypoplastic.

The azoospermic male was 27 years old, and was first studied for infertility. His phenotype was unremarkable, but the karyotype in white blood cells was 45,der(X)t(X;Y)(p22.3;p11.2)/46,t(X;Y)(p22.3;p11.2) signifying the presence of mosaic cell lines, each containing a Y;X translocation.

By use of the modified PRINS method described above, we detected SRY on band Yp11.2-p11.31 [ISCN, 2005] in normal male controls (Fig. 1A) and in cells from the XY woman. Presence of the *SRY* testis-determining gene in a woman with gonadal dysgenesis is not unusual, occurring in more than 80% of cases. It is attributed to perturbation of other genes in the sex-determining pathway or to the pleiotropic effects of genes such as *DAX1 (dosage sensitive sex reversal-adrenal hypoplasia congenital-critical region of the X chromosome, gene 1)*, which block SRY-mediated testis differentiation when they are duplicated (Wachtel, 1998). Further application of PRINS could help to determine whether *DAX1* was in fact duplicated in cells from this patient, or whether failure of male development was due to some other lesion. Thus PRINS can be used to elucidate specific genetic defects.

We observed the SRY signal in Xp22 in the derivative X of the 30-year-old XX male (Fig. 1B) and in the derivative X chromosome of the azoospermic male (by FISH in Fig. 1C; by PRINS in Fig 1D). But the signal was absent in cells from the XX male infant with hypospadias. Absence of signal in the infant with hypospadias is not surprising. Approximately 70% of cases of XX male syndrome are due to X;Y crossover in the germ line of the father. These cases are notable for presence of *SRY*. The other 30%, often characterized by ambiguity of the external genitalia, are attributable to perturbations downstream of *SRY* in the sex-determining pathway. These cases are notable for absence of *SRY*. In that connection, duplication of *SOX9 (SRY-like HMG box 9)* on chromosome 17 has been described in some SRY-negative XX males, consistent with the idea that SRY-mediated induction of *SOX9* is a crucial step in differentiation of the male gonad (reviewed in Wachtel, in press). The question whether *SOX9* is duplicated in this case, could be addressed by application of the methods describe in this chapter.

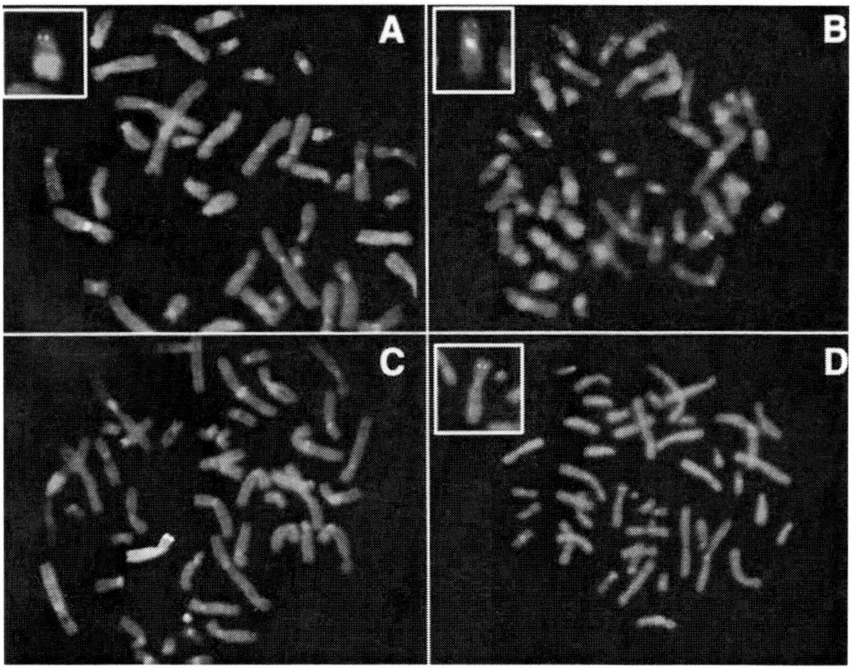

Figure 1. A. PRINS signals (red) for *SRY* in normal male control. The X centromere is stained green by FISH. Similar PRINS and FISH signals were observed in female with 46,XY gonadal dysgenesis
B. PRINS signals (red) for SRY in der(X) of XX male. X centromeres are stained green by FISH
C. The X chromosome is stained green by FISH in cells of the azoospermic male (see text). Y-derived material (yellow) is seen on Xp and the red-stained der(Y)
D. SRY signals (red dots) identified by PRINS on Xp of the azoospermic male

The SOX Family of Genes: Detection of *SOX3*

The *SRY* testis-determining gene is one of a family of genes encoding high mobility group (HMG) transcription factors with a highly conservative DNA binding motif called the *HMG box*. The *SOX9* gene, discussed in the preceding paragraph, is another. Many of the *SOX* genes exhibit tissue-specific patterns of expression in early embryogenesis, consistent with the idea that they play significant roles in tissue-specific induction.

SOX3 (SRY-like HMG box 3)—yet another member of the SOX family—is located on the long arm of the human X chromosome at Xq26q27 (Stevanovic et al., 1993). *SOX3* is 2.4 kb in length and consists of a single exon. The corresponding *Sox3* gene of the mouse, also on the X, is required for neural development and gametogenesis (Raverot et al., 2005). Because the *SOX3* sequence is most closely related to that of *SRY*, and because *SOX3* is X-linked, it has been suggested that *SOX3* represents a primordial X-sequence from which the Y-linked *SRY* gene is derived (Foster and Graves, 1994).

In order to further evaluate the efficacy of the PRINS method in visualizing single copy genes, we tested the cells of a normal 46,XX female for presence of *SOX3*. The fluorescent gene signal was visualized as a doublet—two green dots—representing the two chromatids in the long arm on both X chromosomes. For these studies, identification of the X was facilitated by use of FISH with the DXZ1 probe, after the PRINS reaction, to highlight the X centromere.

Deletion Studies of RBM and DAZ in an Azoospermic Man

Genes governing spermatogenesis have been mapped to the long arm of the human Y chromosome at proximal Yq11.2 in a domain called the *azoospermia factor region (AZF)*. Thus deletions within the AZF region are correlated with various defects of sperm production. AZF has been subdivided into 4 subregions—AZFa, AZFb, AZFc and AZFd. Candidate genes for azoospermia include *RBM (RNA binding motif)* corresponding to AZFb, and *DAZ (deleted in azoospermia)* corresponding to AZFc. Both these genes are expressed in the testes, and both are situated in Yq11.2 (Reijo et al., 1996).

We evaluated cells from a 27-year-old man referred for infertility. He was 172 cm tall and weighed 68.5 kg. He appeared to be in good health but there was a paucity of body hair, and left and right testes were soft and measured 3 cm and 3.5 cm. No sperm cells were found in the seminal fluid (azoospermia). A mosaic karyotype was identified by GTG-banding: 45,der(X)t(X;Y)(p22.3;p11.2)/46,t(X;Y)(p22.3;p11.2). Studies by FISH with dual color paint probes confirmed reciprocal translocations between the short arms of the X and Y chromosomes, and showed absence of Yq11.2, i.e. the part of the Y that contains the AZF region.

PRINS was used to evaluate *SRY, RBM* and *DAZ* in the patient's lymphocytes. *SRY* was located on Xp22, confirming X;Y translocation. The *RBM* and *DAZ* candidate genes were absent, consistent with a role for those elements in spermatogenesis. In aggregate, the data obtained in this study led us to suspect that the derivative Y chromosome was in fact a ring chromosome, formed by Yp-Xp exchange and loss of material from Yq. This could explain loss of the ring in some cells of the patient, and the resulting mosaic karyotype (see discussion in Kadandale et al., 2002).

Gene Deletions in Cancer

The preceding data show too that the PRINS method can be employed gainfully in the detection of microdeletions and subtle chromosome rearrangements. This has been borne out by the use of PRINS in deletion studies of the *RB1 (retinoblastoma 1)* and *p53* tumor suppressor genes in the human (Tharapel and Kadandale, 2002). Rearrangements on chromosomes 13 and 14—in particular regions 13q14 and 17p13—occur frequently in leukemias and lymphomas. But they may not be observable by the standard methods of cytogenetics. In fact, deletions within these regions occur in >50% of cases, yet the karyotypes appear to be normal because the deletions are submicroscopic. In the past, the deletions have been detected by FISH.

Tharapel and Kadandale (2002) applied PRINS in patients with known deletions in these regions and compared their results with those obtained with FISH. They studied 5 patients with deletions in 13q14 (the *RB1* region) and 5 patients with deletions in 17p13 (the *p53* region). The patients in this trial had a normal chromosome 17 and an isochromosome of the 17 long arm, designated i(17q), and were thus monosomic for 17p.

FISH was performed before PRINS. The chromosome preparations from bone marrow cell pellets that were used for FISH were also used for PRINS: Normal controls were processed alongside the patient samples. For each patient and each control, 10 metaphase spreads and 40 interphase nuclei were evaluated in each procedure. FISH and PRINS methods yielded comparable results for normal and deletion patterns in all cases—for both *p53* and *RB1* genes. Although the PRINS signals were smaller than the FISH signals, the authors described the PRINS signals as "distinct and unambiguous."

In such cases, tumors presumably arise as a result of *RB1* or *p53* mutations in the *intact* chromosomes of hematopoietic cells. Having already lost a wild-type copy of the relevant gene due to microdeletion, the affected cells lose suppressor function, and the result is uncontrolled cell division leading to cancer. Neoplastic hematopoietic cells from the patients should accordingly exhibit only one PRINS signal, representing the remaining copy of the *RB1* or *p53* gene. Cells from normal controls should exhibit two signals.

Diagnostic Applications: Gene Deletions in Prader Willi and Angelman Syndromes

Whether a particular chromosome comes from the mother or father can effect significantly the expression of clinical syndromes resulting from genetic lesions on that chromosome. This phenomenon is called *parental imprinting*. An example is the differential appearance of Huntington disease, a degenerative disorder of the central nervous system with progressive dementia and spasmodic movements of the face and limbs, caused by a dominant gene on chromosome 4 (Nussbaum et al., 2004). Patients who inherit the gene from their father show symptoms in adolescence—*early onset* disease. Patients who inherit the gene from their mother show symptoms in middle age—*late onset* disease. A more dramatic example is provided by the differential response to deletions in the proximal long arm of chromosome 15 (q11.2-q13). When the affected chromosome is paternal in origin, the result is the Prader-Willi syndrome. When the affected chromosome is maternal, the result is Angelman syndrome.

Prader-Willi syndrome (PWS) occurs in 1/10,000 to 1/25,000 live births. It is characterized by mild mental retardation, obesity, short stature, small genitalia and small hands and feet. Affected children have a region 15 (q11.2-q13) with genetic information that comes only from the mother because the paternal chromosome 15 has been deleted for this region. The other side of the coin is the phenotypically distinct Angelman syndrome (AS), also called the "happy puppet" syndrome. Children with this condition are severely retarded and cannot speak. They are characterized by spastic "puppet-like" movements, easily provoked laughter, microcephaly and extruding tongue. Although the chromosomal region deleted in this condition is the same one that is deleted in PWS (15q11.2-q13), it is deleted here from the *maternal* copy of chromosome 15. Thus patients with AS have genetic information in 15 (q11.2-q13) deriving only from the father. In some cases, the disorder is found in children who have inherited both chromosomes 15 from the father and thereby lack a maternal 15 *(uniparental disomy)* (Nicholls, 1993).

For evaluation of patients with PWS or AS, we used primer sequences specific for the *SNRPN (small nuclear ribonucleoprotein polypeptide N)* and *GABRB3 (gamma-aminobutyric-acid receptor beta-3 subunit precursor)* genes, both thought to reside normally in the deleted regions of chromosome 15. PRINS primer sequences were derived from published sequences for these genes. We studied 4 males and 4 females with PWS, a male and a female with AS (Tharapel et al., 2002). Three normal males and 2 normal females served as controls. In all cases, venous blood was collected in sodium heparin.

Phytohemagglutinin-stimulated lymphocyte cultures were harvested, and slides prepared according to conventional protocols. Chromosome preparations of the patients were exposed to Wright stain after treatment with a dilute solution of pancreatin, for G-banding; 20 cells were studied for each patient. For FISH, we obtained *SNRPN* probes from Vysis Inc., Downer's Grove, IL. The *SNRPN* locus is generally deleted in patients with PWS and AS. We analyzed 20 metaphase spreads and 80 interphase nuclei from each patient and compared the results with those obtained after PRINS.

In all cases, our PRINS results were in full agreement with the results obtained by FISH (Table 1). Thus, among the 5 controls, a normal staining pattern was observed after PRINS in >90% cells studied for *SNRPN* and in >90% studied for *GABRB3*. And the corresponding pattern was observed after FISH with probes for the same genes. But among 4 patients with PWS and one with AS, a deletion was observed after PRINS in >80% of cells studied for *SNRPN* and *GABRB3*. And the corresponding pattern was observed for the same patients after FISH with a probe for *SNRPN* (the *GABRB3* locus was not studied with FISH in these patients). Normal staining patterns were observed by FISH and PRINS in the other 4 patients with PWS and one with AS (Table 2). Normal patterns could be attributed to uniparental disomy in these patients, or occasionally, to ambiguity of diagnosis.

Table 2. Analysis by PRINS and FISH of 15q11.2-q13 microdeletions

Subjects	N	Test	Probe/ locus	Labeling pattern	In percent cells **
Controls	5	PRINS	SNRPN	Normal	>90
			GABRB3	Normal	>90
PWS/AS	5	PRINS	SNRPN	Normal	>90
			GABRB3	Normal	>90
PWS/AS *	5	PRINS	SNRPN	Deleted	>80
			GABRB3	Deleted	>80
Controls	5	FISH	SNRPN	Normal	>95
PWS/AS	5	FISH	SNRPN	Normal	>95
PWS/AS *	5	FISH	SNRPN	Deleted	>90

* Four patients with PWS; one with AS

** Scores of less than 100% are attributed to poor nuclear or chromosomal morphology

100 cells tested for each subject

Diagnostic Applications: Gene Deletions in DiGeorge/Velocardiofacial Syndrome

The DiGeorge syndrome (DGS) is characterized by congenital absence of the thymus and parathyroid glands with multiple associated abnormalities including cardiac and aortic malformations, characteristic facial appearance with malformed ears, micrognathia, cleft palate, short philtrum etc. In the absence of an effective T-cell population, there is an increased susceptibility to viral and fungal infections. Most patients with DGS have a hemizygous deletion in chromosome 22(q11.2) which subsumes these loci: *DGCR2 (DiGeorge syndrome critical region gene 2)* and *TUPLE1 (TUP like enhancer of split gene 1) and N25.* The incidence of children with the deletion is about 1/4000 live births. The DiGeorge critical region spans 2 mb and is deleted in about 90% of DiGeorge patients. The *N25* locus (D22S75) is situated in the proximal end of the DiGeorge critical region, whereas the *TUPLE1* gene is located within a minimum critical region of 480-575 kb.

The Velocardiofacial syndrome (VCFS) is associated with deletions in the same region of chromosome 22 that DiGeorge syndrome (DGS) is. These patients sometimes show extended deletions: 22(q11.2-q12.3). Patients with this condition commonly exhibit cleft palate, cardiac malformations, abnormal facies, learning disabilities, short stature, immune defficiency, hypoparathyroidism and other malformations. The incidence has been reported as greater than 1/1000, although VCFS and DGS appear to have a similar etiology and thus may represent alternative molecular lesions within the same critical region of chromosome 22.

We studied 3 females and 4 males with DGS, all of whom had the characteristic malformations of derivatives of the third and fourth paryngeal pouches: hypoplastic thymus and parathyroid, diverse abnormalities of the aortic arch, and mild dysmorphic features. We also studied 2 females and 1 male with VCFS. Those patients were notable for short stature, elongated facies, submucosal cleft palate, tetralogy of Fallot, long thin hands and feet, and mild to moderate mental retardation.

As in the previous cases, we collected venous blood in heparinized syringes, and analyzed 20 metaphase spreads and 80 interphase nuclei from each patient. For FISH, we used the N25 cosmid probe (D22S75) which hybridizes with sequences in the proximal end of the DiGeorge critical region. Absence of N25 signal would be expected in cases of microdeletion within 22q11.2. In fact the deletions in VCFS and DGS are usually indistinguishable at the chromosomal level, consistent with a similar etiology in these conditions.

For PRINS, we use primers derived from published sequences of the *DGCR2* and *TUPLE1* genes (Table 1). The results of our tests in 10 patients with DGS/VCFS are shown in Table 3. By use of PRINS, we identified deletions in >80% of cells from 5 patients with DGS and 1 patient with VCFS. The results were confirmed by FISH. Deletions were not found in controls or in 4 of the patients. Analyses by PRINS and FISH gave similar results in all cases (Tharapel et al., 2002).

Table 3. Analysis by PRINS and FISH of 22q11.2 microdeletions

Subjects	N	Test	Probe/ locus	Labeling pattern	In percent cells **
Controls	5	PRINS	DGCR2/TUPLE1	Normal	>90
DGS/VCFS	4	PRINS	DGCR2/TUPLE1	Normal	>90
DGS/VCFS*	6	PRINS	DGCR2/TUPLE1	Deleted	>80
Controls	5	FISH	N25	Normal	>95
DGS/VCFS	4	FISH	N25	Normal	>95
DGS/VCFS*	6	FISH	N25	Deleted	>90

* Five patients with DGS; one with VCFS

** Scores of less than 100% are attributed to poor nuclear or chromosomal morphology

100 cells tested for each subject

Conclusion

The studies reviewed above suggest potential applications of PRINS in the evaluation of individual genes, gene duplications and microdeletions. In fact, with further development and adaptation, the PRINS method might have a fundamental impact on medical genetics; if the method could be applied routinely, physical mapping of individual loci would be possible in most laboratories concerned with the study of human chromosomes and the diagnosis of genetic disease.

Certain difficulties remain to be overcome. The PRINS signal is small compared with the FISH signal. This problem may be alleviated by the use of more than one primer for single loci, and by combining annealing and extension into a single step (we have used as many as 5 primers for a single locus; Tharapel and Kadandale, 2002). Moreover, primer concentrations must be optimized for optimal hybridization signal. This is a critical step in the PRINS reaction, as improper concentration of primer can compromise the result. Background staining can be minimized by increasing stringency of washes, and low signal can be reduced by optimizing concentration of other reagents and by increasing reaction times (see Notes in Tharapel and Wachtel, in press).

Despite these shortcomings, we look forward to more widespread use of the PRINS method and to broader application of PRINS in the study of individual loci and gene deletions in the diagnosis of various cancers and microdeletions, and of genetic disorders in general.

References

Cinti, C., Santi, S., and Maraldi, N.M. (1993) Localization of single copy gene by PRINS technique. *Nucleic Acids Res. 21,* 5799-5800.

Foster, J.W., and Graves, J.A. (1994). An SRY-related sequence on the marsupial X chromosome: implications for the evolution of the mammalian testis-determining gene. *Proc. Nat. Acad. Sci. U.S.A. 91,* 1927-1931.

Gosden, J., Hanratty, D., Starling, J., Fantes, J., Mitchell, A., and Porteous, D. (1991). Oligonucleotide-primed in situ DNA synthesis (PRINS): a method for chromosome mapping, banding, and investigation of sequence organization. *Cytogenet. Cell Genet. 57,* 100-104.

ICSN 2005. An International System for Human Cytogenetic Nomenclature (2005). (2005). L.G.Shaffer, and N. Tommerup (Eds.). S. Karger, Basel.

Kadandale, J.S., Wachtel, S.S., Tunca, Y., Martens, P.R., Wilroy, R.S., and Tharapel, A.T. (2002). Deletion of RBM and DAZ in azoospermia: evaluation by PRINS. *Amer. J. Med. Genet. 107,* 105-108.

Kadandale, J.S., Wachtel, S.S., Tunca, Y., Wilroy, R.S. Jr., Martens, P.R., and Tharapel, A.T. (2000). Location of SRY by primed in situ labeling in XX and XY sex reversal. *Amer. J. Med. Genet. 95,* 71-74.

Koch, J., Hindkjaer, J., Kolvraa, S., and Blund, L. (1995). Construction of a panel of chromosome-specific oligonucleotide probes (PRINS-primers) useful for the identification of individual human chromosomes in situ. *Cytogenet. Cell Genet. 71,* 142-147.

Koch, J.E., Kolvraa, S., Gregerson, N., Petersen, K.B., and Bolund, L. (1989). Oligonucleotide-priming methods for the chromosome-specific labeling of alpha satellite DNA in situ. *Chromosoma 98,* 259-265.

Nicholls, R.D. (1993). Genomic imprinting and candidate genes in the Prader-Willi and Angelman syndromes. *Curr. Opin. Genet. Dev. 3,* 445-446.

Nussbaum, R., Thompson, M.W., McInnes, R.K., and Willard, H.F. (2004). *Thompson and Thompson Genetics in Medicine.* Philadelphia: W.B. Saunders.

Pellestor, F., Girardet, A., Andreo, B., and Charlieu, J.P. (1994). A polymorphic alpha satellite sequence specific for human chromosomes 13 detected by oligonucleotide primed in situ labeling (PRINS). Hum. Genet. 94, 346-348.

Pellestor, F., Girardet, A., Lefort, G., Andreo, B., and Charlieu, J.P. (1995). Use of primed in situ labeling (PRINS) technique for a rapid detection of chromosomes 13, 16, 18, 21, X and Y. *Hum. Genet. 95,* 12-17.

Raverot, G., Weiss, J., Park, S.Y., Hurley, L., and Jameson, J.L. (2005). Sox3 expression in undifferentiated spermatogonia is required for the progression of spermatogenesis. *Dev. Biol. 283,* 215-225.

Reijo, R., Alagappan, R.K., Patrizio, P., and Page, D.C. (1996). Severe oligospermia resulting from deletions of azoospermia factor gene on Y chromosome. *Lancet 347,* 1290-1291.

Sinclair, A.H., Berta, P., Palmer, M.S., Hawkins, J.R., Griffiths, B.L., Smith, M.J., Foster, J.W., Frischauf, A.M., Lovell-Badge, R., and Goodfellow, P.N. (1990). A gene from the human sex-determining region encodes a protein with homology to a DNA binding motif. *Nature 346,* 240-244.

Stevanovic, M., Lovell-Badge, R., Collignon, J., and Goodfellow, P.N. (1993). SOX3 is an X-linked gene related to SRY. *Hum. Molec. Genet. 2,* 2013-2018.

Tharapel, A.T. and Wachtel, S.S. (In press). PRINS for detection of Gene deletions in cancer, In F. Pellestor (Ed.), *PRINS and in situ PCR.* Totowa, New Jersey: Humana Press. (In press).

Tharapel, A.T.,Kadandale, J.S., Martens, P.R., Wachtel, S.S., and Wilroy, R.S., Jr. (2002). Prader Willi/Angelman and DiGeorge/Velocardiofacial syndrome deletions: Diagnosis by primed in situ labeling (PRINS). *Amer. J. Med. Genet. 107,* 119-122.

Tharapel S.A., and Kadandale, J.S. (2002). Primed in situ labeling (PRINS) for evaluation of gene deletions in cancer. *Amer. J. Med. Genet. 107,* 123-126.

Wachtel, S.S. (1998). X-linked sex-reversing genes. *Cytogenet. Cell Genet. 80,* 222-225.

Wachtel, S.S. SRY—sex reversal, In D. Ganten, and K. Ruckpaul (Eds.), *Encyclopedic Reference of Genomics and Proteomics in Molecular Medicine.* Berlin, Heidelberg, New York: Springer. (In press).

In: PRINS and PNA Technologies in Chromosomal Investigations ISBN 1-60021-413-4
Editor: Franck Pellestor, pp. 37-52

Chapter 4

Identification of all Fetal Nucleated Cells in Maternal Blood Using Prins Technique

Kada Krabchi, Macoura Gadji, Ju Yan and Régen Drouin*
Service of Genetics, Department of Paediatrics,
Faculty of Medicine and Health Sciences, Université de Sherbrooke, Sherbrooke, Quebec, Canada

Abstract

Fetal nucleated cells (FNCs) exist in maternal blood during pregnancy. Many groups explore the possibility of developing a reliable non-invasive prenatal diagnosis of chromosomal aberrations by detecting FNCs circulating in maternal blood. The total number of FNCs was mostly quantified using markers present in every fetal cell such as sex chromosomes (X and Y) and chromosomes 21, 18, and 13 involved in aneuploidies by molecular cytogenetic techniques. Fluorescence *in situ* hybridization (FISH) and PRimed *IN Situ* labeling (PRINS) are two sensitive molecular cytogenetic techniques that have empowered the conclusive demonstration of the presence of the few FNCs circulating in maternal blood. In order to ascertain the exact number of FNCs per unit volume of maternal blood, we performed both techniques (separately or in combination) on peripheral blood samples obtained from 11 pregnant women at different stages of mid-trimester. PRINS reaction is a non-isotopic *in situ* labeling method used for a rapid and efficient interphase sex chromosome and autosome aneuploidy detection approach. This method is based on specific oligonucleotide primers derived from chromosome specific subsets of repeated DNA sequences, followed by subsequent primer extension using *Taq* DNA polymerase. Using specific primers for chromosomes X, Y and 18, we performed the PRINS reaction, followed by specific antibody-fluorophore complex detection. Our results displayed an average of 3 FNCs per mL (or a ratio of ~ 1 FNC/ 2 to

* Correspondence to: Régen Drouin, Service of Genetics, Department of Paediatrics, Faculty of Medicine and Health Sciences, Université de Sherbrooke, 3001, 12th Avenue North, Sherbrooke, Quebec, Canada J1H 5N4; Tel.: (819) 820-6827; Fax:(819) 564-5217; Email: Regen.Drouin@USherbrooke.ca

$3x10^6$ maternal cells) of maternal blood in euploid pregnancies. This number was 4-fold higher in aneuploid pregnancies (e.g. trisomy 18 and trisomy 21). We think that one of the most efficient molecular cytogenetic tool applied to detect FNCs is PRINS labeling. The results of our investigation suggest that PRINS might be a powerful tool, more specific, considerably faster and less expensive than conventional FISH. Moreover this cost-effective method could be useful in non-invasive prenatal diagnosis.

Key words: Primed in situ labeling, fetal nucleated cells, maternal blood, non-invasive prenatal diagnosis

Introduction

Fetal cells in maternal blood circulation was first confirmed by Georg Schmorl (Leipzig - Germany, 1893) who detected syncytiotrophoblastocytes (multinucleated syncytial giant cells) trapped in the uterine veins and lung capillaries of 14 of 17 pregnant women who died of eclampsia [1]. Since then, several manuscripts have reported the presence of fetal cells in the maternal blood and other tissues. Evidence for their existence in maternal blood was first presented in the late 1950s through the identification of trophoblastocytes, which are derived from the placenta [2]. In the following years, Y-chromosome-containing cells were found in the blood of pregnant women carrying a male fetus [3,4]. Numerous research teams have investigated the nature of the fetal cells in the maternal circulation. Controversy still surrounds the duration of FNCs survival in the maternal circulation, the specificity of various FNC characteristics (morphology, surface antigens, protein content, etc.) in their differentiation from maternal cells, FNCs quantity, and the time of FNCs appearance and disappearance from the maternal blood. The FNC cells type identified in the maternal circulation include trophoblasts, nucleated erythrocytes, lymphocytes, granulocytes, and platelets [5]. All, except platelets, contain genomic DNA and thus could theoretically be used for the prenatal diagnosis of chromosomal abnormalities.

Trophoblast sprouts cross through the placental barrier extensively in early pregnancy but they are cleared rapidly, most being trapped by the pulmonary circulation [1]. Taking into account the unexpected occurrence of a confined placental mosaicism, the possibility of a genetic mosaicism of this type of fetal cells must be considered. Cytotrophoblast cells have the potential to be the first fetal cells to appear in maternal blood considering their invasion of maternal spiral arterioles that occurs by the sixth to the eighth weeks of gestation. They are theoretically the most attractive candidates for the purification of fetal cells from maternal cells because of their unique morphology. Cytotrophoblasts identified in maternal blood are probably mononuclear cytotrophoblastic cells [6-8]. Syncytiotrophoblast cells could survive for a long time in the maternal circulation because they do not express HLA class I and class II antigens [9]. However, due to their size, syncytiotrophoblast cells are probably cleared rapidly by the reticuloendothelial system.

Nucleated red blood cells (NRBCs) are likely the most prevalent fetal cell type in the maternal circulation [10-13]. They are mononuclear and relatively frequent in first trimester fetal blood. However, the majority of the NRBCs found in maternal blood are of maternal origin [14-16] and thus must be distinguished from their fetal counterparts. Fetal hemoglobin

(HbF) has been used for to distinguish these cells but it is also present in maternal erythroblasts [9,12]. Embryonic hemoglobins [17,18] have also been used but their expression decreases rapidly during pregnancy [19]. The limited lifespan of NRBCs is about 90-120 days [20] thereby, rendering them unlikely to persist across pregnancies [21]. Recently, Cha et al. developed a scoring system to objectively distinguish fetal NRBCs from maternal cells [22].

Fetal lymphocytes may or may not exist in maternal blood. The transfer of fetal lymphocytes in maternal circulation could explain the development of HLA antibodies in pregnant women. However, the presence of HLA antigens should limit the survival of these cells in maternal circulation unless a pregnancy-induced tolerance is present. Using FISH with a Y-chromosome-specific DNA probe in maternal blood without enrichment of fetal cells, Hamada et al. [23] failed to identify a signal in 4 out of 16 pregnancies with a male child after examining 100,000 nucleated cells (not just lymphocytes). With their lifespan of several years, fetal lymphocytes, if they are tolerated by the mother's immune system, could remain in the maternal circulation after delivery. Several reports have described the presence of such fetal lymphocytes (or of fetal cells without identifying the cell type) [24-26], but controversies about the specificity of the techniques used to identify male cells in the maternal circulation have not yet been resolved. A lack of specificity of the fluorescent probe can lead to identification of XY mitosis in newborn girls [27] or a high percentage of Y-positive cells in non-pregnant women [28]. Other potential explanations for the finding of Y-specific DNA sequences in pregnancies with a female fetus are the presence of a vanishing-twin pregnancy or previous maternal blood transfusion from a male donor [5,29]. Fetal progenitor cells can persist in maternal circulation for many years after delivery [26,30] resulting in the constitution of a fetal microchimerism. Fetal microchimerism may be etiologically associated with maternal autoimmune diseases such as systemic sclerosis [31,32].

Fetal granulocytes exist in maternal blood [33,34]. The high proportion of fetal cells reported in the latter study which used *in situ* hybridization (FISH), has not been replicated by other groups possibly reflecting a lack of specificity of the probe. Their lifespan in maternal circulation is presumably very short [35].

Incontrovertible evidence for the presence of FNCs came with the development of sensitive cytogenetic techniques of genetic analysis, such as FISH [36-38] and PRINS [38-40]. Both techniques target specific DNA sequences and identify them using fluorescent-labeled probes. Additionally, the polymerase chain reaction (PCR) which amplifies very small amounts of chromosome DNA sequences specific to the fetus can identify dominant genetic traits or fetal gender [41-43].

Today a consensus exists that fetal cells are rare but universally present in peripheral blood of pregnant women. Fetal blood cells can be recovered and isolated from maternal circulation by different techniques that obviate the risks to the developing fetus associated with invasive procedures (amniocentesis and chorionic villus sampling).

Molecular cytogenetic methods such as FISH and PRINS techniques significantly advanced our knowledge of FNCs in maternal blood. The technical principles underlying PRINS reaction are briefly described hereunder. Primed *in situ* labeling (PRINS) is a cytological technique that combines the advantages of both PCR and FISH [44]. It is based on the *in situ* annealing of specific and unlabeled DNA primers and their subsequent *in situ* extension along chromosomes by a heat stable DNA polymerase in the presence of labeled dNTP analogs. During the primer extension, the DNA polymerase incorporates different nucleotides, dATP, dGTP, dCTP and dTTP, one of which has been chemically modified with an hapten, such as biotin- or digoxigenin-dUTP, is added to the reaction. These modified nucleotides are incorporated in the newly synthesized DNA strand during the primer extension and subsequently revealed in individual cells using fluorochrome labeled ligands (antibodies, avidin, etc.). For PRINS, the high specificity and sensitivity of *in situ* primer extension are obtained since the primers used are highly specific for each chromosome. A panel of specific primers for most human chromosomes have been described [45]. For FISH analysis, specific α-satellite repetitive DNA probes are generally hybridized. Chromosomes 13 and 21 α-satellite repetitive DNA sequences show high homology, requiring the use of specific single-copy sequences.

In this chapter, we will describe the PRINS technique and its application to the detection of a rare event represented by FNCs in maternal circulation in normal male singleton pregnancies and in trisomic 18 and 21 pregnancies.

Materials and Methods

Clinical Material

Following Institutional Review Board approval, genetic counseling and informed written consent, a sample of peripheral venous blood was taken from each of 11 pregnant women carrying euploid and aneuploid male fetuses. All were from a spontaneous singleton pregnancy. This cohort was represented by 2 normal male fetuses, 5 male fetuses with trisomy 21 and 4 male fetuses with trisomy 18 (see Table 1). All whole-blood specimens were collected at different stages of mid-trimester at 1 to 4 weeks after the amniocentesis and before any medication was given for termination of pregnancy.

Harvesting Method

Three milliliters of peripheral venous blood from each pregnant woman were taken and collected in a Vacutainer sodium heparin tube (Becton Dickinson VACUTAINER Systems, Franklin Lakes, NJ). Blood samples were transferred to the laboratory as soon as possible (between half an hour to 4 hours).

The method of cell harvesting and slide spreading were performed as follow: 3 mL of sodium chloride (0.9%) solution were added into a VACUTAINER® K_2 EDTA 3.6 mg (EthyleneDiamineTetraAcetic acid) lavender top tube.

Table 1. Clinical data and number of male FNCs per mL of maternal blood observed in cases of euploid and aneuploid pregnancies

Cases	GnPnAn*	M.A.	S.S.	F.N.C.	G.A.	Ratio
XY	$G_3P_1A_1$	29	46	4 (2 mL)	$19^{6/7}$	2
XY	$G_1P_0A_0$	18	18	5 (1 mL)	21	5
T18	$G_2P_1A_0$	38	23	20 (1 mL)	17	20
T18	$G_3P_0A_2$	41	26	11 (1 mL)	$18^{3/7}$	11
T18	$G_1P_0A_0$	26	26	7 (1 mL)	$21^{3/7}$	7
T18	$G_2P_0A_1$	28	25	4 (1 mL)	18	4
T21	$G_4P_1A_2$	42	42	32 (2 mL)	$18^{3/7}$	16
T21	$G_3P_1A_1$	32	42	24 (2 mL)	18	12
T21	$G_3P_2A_0$	37	31	14 (1 mL)	21	14
T21	$G_5P_2A_2$	42	25	11 (1 mL)	$15^{4/7}$	11
T21	$G_1P_0A_0$	39	20	8 (1 mL)	21	8

* GnPnAn (G: Gestation, P: Parity, A: Abortion)

M.A.: maternal age; M.C.: number of male fetal nucleated cells detected and volume of maternal blood analyzed; G.A.: gestational age; Ratio: number of FNCs per mL of maternal blood; S.S: slides scored; XY: male euploid pregnancies; T18 & T21: male aneuploid pregnancies (4 trisomy 18 and 5 trisomy 21)

**The 2 male euploid cases and 4 male trisomy 18 cases were analyzed using dual-color PRINS technique and the 5 male trisomy 21 cases were analyzed using PRINS combined to FISH technique

Three mL of fresh whole blood was mixed with 3 mL of sodium chloride 0.9% solution, taken from a *VACUTAINER*® K_2 EDTA tube. Six equal 1-mL volumes were aliquoted into 6 Falcon polypropylene plain tubes (15 mL). After washing the cells with HBSS (Hank's Balanced Salt Solution) twice, the cells were centrifuged at 1500 RPM (ca. 500g, ICE centrifuge, HN-SII model A Division of Damon, USA) for 7 minutes and the platelet-rich supernatant was discarded. The cells were resuspended in a hypotonic solution (0.075 *M* KCl) and incubated at 37°C for 10 to 15 minutes. After cytocentrifugation for 5 minutes at 2000 RPM (ca. 667g), the cells were quickly prefixed in 5% glacial acetic acid solution before being fixed several times with freshly prepared precooled Carnoy's fixative (3v:1v, methanol : glacial acetic acid). The pellet was resuspended in 1 mL of fresh Carnoy. The cell suspensions were pooled in 3 microtubes (the volume equivalence of blood per microtube is 1 mL). These suspensions were stored in Carnoy at –20°C for up to many months. 15 μL of fixed cell suspension was spread onto cleaned slides in a modified Thermotron environmental control unit at an optimal temperature (25° C) and humidity (36%). Slides were then immersed and agitated in methacarn solution (6v:3v:1v, methanol:chloroform:glacial acetic acid) for 20-30 seconds. Slides were then put on racks and in the Thermotron chamber to dry. They were used immediately or stored at –20°C for several weeks before use [38,40].

Protocols Performed in this Study

The dual-color PRINS used in this study was inspired by the method described by Yan et al. [46,47]. It is performed to detect two different chromosomes by omitting the 3'-end blocking step. Omission of the blocking step never affected the correct identification of two chromosomes because the signals from the second PRINS reaction site always show a pure single original color.

Two different hybridization protocols were used:

Protocol 1, dual color-PRINS: the reaction is initiated by adding the first mix solution (50 μL) composed for each slide of: 5 μL of 10x Polymerase Chain Reaction (PCR) buffer (Roche Molecular Biomedicals), 4 μL of each 2.5 μ*M* dATP, dCTP and dGTP, 4 μL of 0.25 μM dTTP, 2 μL of 2.5 μ*M* primer solution, 1 μL of 1 m*M* digoxigenin-11-dUTP or biotin-11-dUTP, 2,5 μL of glycerol, 23 μL of distilled water, and 0.5 μL of 2.5 U Taq DNA polymerase, containing a specific X-chromosome primer (Xc: 5-GTT CAG CTC TGT GAG TGA AA-3') [48] and one of the labels onto a slide. For X chromosome detection, the PRINS program should be annealed at 65°C for 10 min, with an extension at 72°C for 10 minutes.

For Y- chromosome primers (Y: D599: 5'-TGG GCT GGA TGA GAA AGG AAT CGA AAC-3' and D600: 5'-TCC ATT CGA CAT TTT TTT CGA GAA-3') [49] annealing at 56°C for 10 minutes and an extension at 72°C for 10 min is advised.

For specimens with male trisomy 18, the first PRINS reaction used the centromeric specific sequence of chromosome 18 (18c: 5′-ATG TGT GTC CTC AAC TAA-AG -3′) at an optimal annealing and extension temperatures (62.5°C) [48] for 10 minutes each, followed by the second PRINS reaction using the two Y-chromosome centromeric specific sequences.

Slides are covered with a cover slip and placed onto the flat block of a thermocycler (PTC-100, MJ Research).

After completion of the first PRINS reaction, the slides are briefly washed in 1X PBS solution for 2 minutes. The second reaction solution contains the Y specific primers for the chromosome Y and an alternately labeled dUTP on the same slide. The use of two primers for the same locus of chromosome Y increases the efficiency of PRINS reaction. No denaturation is required after the first PRINS reaction because the DNA remains denatured throughout the PRINS incubation.

The slide is washed in buffer for 5 minutes with gentle agitation and mounted with cover slips and sealed with rubber cement (vulcanizing solution).

Protocol 2: involves a combination of PRINS followed by FISH [44,48]. PRINS is with the previously described (protocol 1) specific centromeric Y-chromosome primers. The subsequent FISH technique uses a locus specific fluorescent labeled probe to chromosome 21 (LSI 21 SpectrumOrange, Vysis Inc, Downers Grove, IL. USA). This second protocol requires that DNA on slides must be strictly denatured with formamide 70%/2xSSC in a water bath with accurate temperature control at 72°C for 2 minutes. The annealing and extension steps of the Y chromosome by PRINS were set as performed in dual-color PRINS. The LSI-21 FISH probe was added after denaturing at 72°C for 5 minutes. Slides were mounted with cover slips and sealed them with rubber cement (vulcanizing solution). These slides were then placed in a humid chamber at 37°C overnight for probe hybridization and washed on the following day. The specific hybridized PRINS Y- probe was detected by an anti-digoxigenin antibody labeled with fluorescein.

Detection

Slides were mounted with 100 μL of blocking buffer and a cover slip and left at room temperature for 5 minutes. The cover slips were removed, and blocking buffer was drained from the slide. 100 μL of a mix of detection solution (1% avidin-fluorescein and 1% anti-dig-rhodamine diluted with blocking buffer or 1% Avidin-rhodamine and 1% Anti-digoxigenin-fluorescein diluted with blocking buffer) was applied to the slide. Either of the two detection mixes above can be used according to the selected labeling protocol for the detection of two chromosome targets.

A new cover slip was mounted and the slide was incubated in a moist chamber at 37°C for 30 minutes.

The cover slips were removed gently by tilting the glass slide, and letting the cover slip slide off.

Slides were washed three times in washing buffer at room temperature for 5 minutes each with gentle agitation to remove excess antibody.

After the slide air-dried in darkness, 10-20 μL DAPI (4',6-diamidino-2-phenylindole) counterstaining solution was applied and slides were mounted under a cover slip in a solution containing p-phenylenediamine (Sigma Chemical, St-Louis, MO, USA) to inhibit quenching.

Microscopic Observation

The fetal and maternal cells showing representative signals are recorded using a Compulog IMAC-CCD S30 video camera module and analyzed using the *in situ* imaging system mounted on the fluorescence microscope (Olympus BX60 microscope, Metasystems, Belmont, MA).

Given the brightness of the signals, the slides are scanned at 400 x magnification air immersion but every cell showing signals pattern different of that harbored by maternal nuclei are examined at x1000 magnification oil immersion.

In euploid pregnancies, male fetal cells are identified by the presence of two fluorescent signals of different colors (*see* Fig.1) corresponding to the X- and Y-chromosomes, respectively. The rehybridization of both chromosomes withy reverse color labeling confirmed the authenticity of FNC [38].

For aneuploid cases, different approaches were used to confirm the fetal origin. Male aneuploid fetal cells were identified by the presence of Y-chromosome signal and three signals from chromosome 21 or from chromosome 18. In nuclei showing false positive signals, for example, three hybridization signals corresponding to chromosome 21 and no Y-chromosome suggesting a false-positive 21 signal, the probes were stripped and rehybridized using different probes labeled with different fluorochromes. Very rare false-positive nuclei (either one signal corresponding to the chromosome Y with two signals corresponding to chromosome 18 or three signals corresponding to chromosome 18 and no signal corresponding to the chromosome Y) were observed. In order to eliminate false positive cells, the signals of the first PRINS reaction were stripped, and the slides were rehybridized using FISH probes labeled with fluorochromes of different colors. All false-positive signals were confirmed as false-positive signal using this approach.

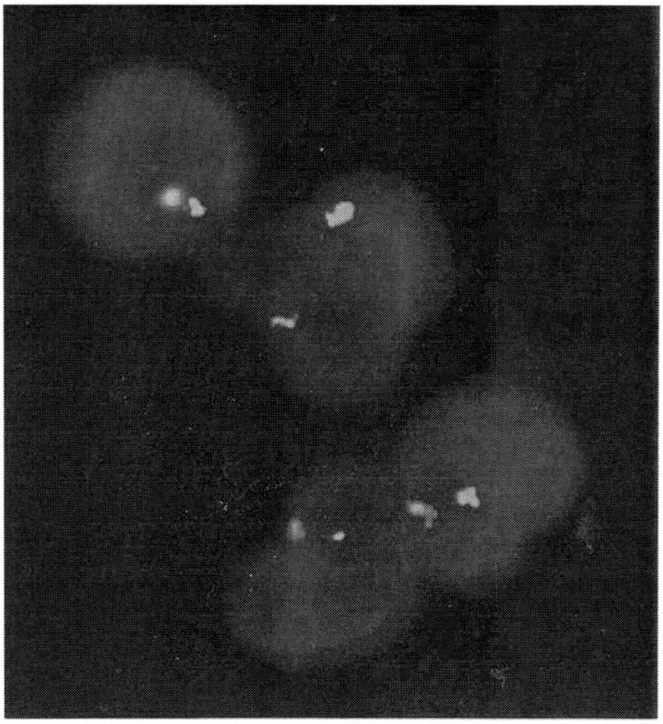

Figure 1. Male fetal cell from maternal peripheral blood detected by dual-color PRINS using X- and Y-specific primers labeled in red (rhodamine) and green (FITC: fluorescein isothiocyanate), respectively. Image capturing was performed at x1000 magnification

Results

The prior obstetrical history of each pregnant woman and the results of our study are shown in Table 1.

A volume of 15 μL of fixed nuclear suspension was spread on each slide that represented approximately 200,000 – 300,000 cells. An average of 30 microscope slides were examined for each pregnant woman. This was equivalent to approximately 6–9 x 10^6 nuclei per sample. Each slide required an average of 1-2 hour scanning time including the localization, and image capturing of each positive FNC found.

Medium-sized male mononucleated cells showing only a Y-positive signal or both Y-positive signal and three signals of either chromosome 18 or chromosome 21 confirming the fetal origin were detected and scored. The vast majority of the nuclei appeared round or oval in shape. We could not accurately distinguish the FNC cell type. However, cell morphology was consistent with a high proportion of the detected fetal cells being trophoblastic cells or erythroblast cells.

Our results were reported [38,50,51]. When twelve normal male pregnancies between the 18th and 22nd weeks of gestation were analyzed [38], there was an average yield of 2 to 6 fetal cell/mL of maternal blood. Among them, the 2 cases studied using PRINS technique showed 2 FNCs in the first euploid pregnant woman and 5 FNCs in the second woman.

Our two more recent studies showed that frequency of FNCs in maternal blood increased 3- to 5-fold in aneuploid singleton pregnancies [50,51]. The results of the present study showed in the 5 cases of trisomy 21 studied with PRINS combined to FISH technique a range of 8 to 16 FNCs per mL; in the 4 cases of trisomy 18 studied with PRINS technique, there was a range of 4 to 20 FNCs per mL. The higher FNC frequency of aneuploid pregnancies, 2 to 5-fold higher than in euploid pregnancies, could be due to early impairment of the feto-maternal barrier.

Discussion

We determined that the specificity and sensitivity of the PRINS technique makes it a valid alternative to the FISH technique for the identification of rare (a ratio of ~ 1 FNC/ 2 to $3x10^6$ maternal cells) fetal nucleated cells (FNCs) in maternal blood. The rare FNCs are identified as those having a Y-chromosome and, in the cases of trisomy for chromosomes 13, 18, and 21, the Y-positive FNCs can be confirmed as FNCs by using autosome specific probes.

Obtaining fetal material from maternal blood avoids the risks associated with invasive prenatal procedures, e.g., amniocentesis, chorionic villus sampling and fetal blood sampling but requires enhanced sensitivity and specificity to detect rare FNCs. The frequency of fetal cells among circulating maternal mononuclear cells has not been established accurately for different stages of pregnancy and many aspects of the fetal cell research are controversial or and remain to be ascertained. Some harvesting procedures utilize cell surface antigens or cell density to enrich maternal blood for FNCs [37,52-54]. Some fetal cells, while in the maternal circulation, are fragile, engaged in the programmed cell death or apoptotic process [55,56]. These FNCs can be lost during harsh harvesting and enriching procedures. We quickly transfer blood to the laboratory and harvested all cells rapidly to avoid loss of fragile cells. This harvesting method uniquely allows identification of all different FNC cell types trafficking in the maternal blood during pregnancy.

The PRINS detection method was designed to detect all types of FNCs using specific cell-type-independent markers, such as for the Y-chromosome solely or both the Y-chromosome and a chromosome representing a type of aneuploidy (eg. trisomy 18 or trisomy 21). The Y-chromosome (Figure 1) and the aneuploid chromosomes are to our knowledge the only universal markers (present in every single fetal nucleated cell, Figures 2a and 2b) that can be used to specifically detect all categories of FNCs. To quantify fetal cells in this study (i.e euploid and aneuploid pregnancies), we always considered one or two of these universal markers. Oligonucleotide primers synthesized from the consensus sequence of a repetitive DNA family can be used to achieve PRINS detection of the target sequences in the chromatin

[45,48,57] for the Y and chromosome 18. Four cases of trisomy 18 (cases 3, 4, 5, and 6) were analyzed using dual-color PRINS. The rare false-positive nuclei were an average of 2 cells/mL of maternal blood and were subsequently identified as false positives using rehybridization of FISH probes labeled with fluorochromes of different colors.

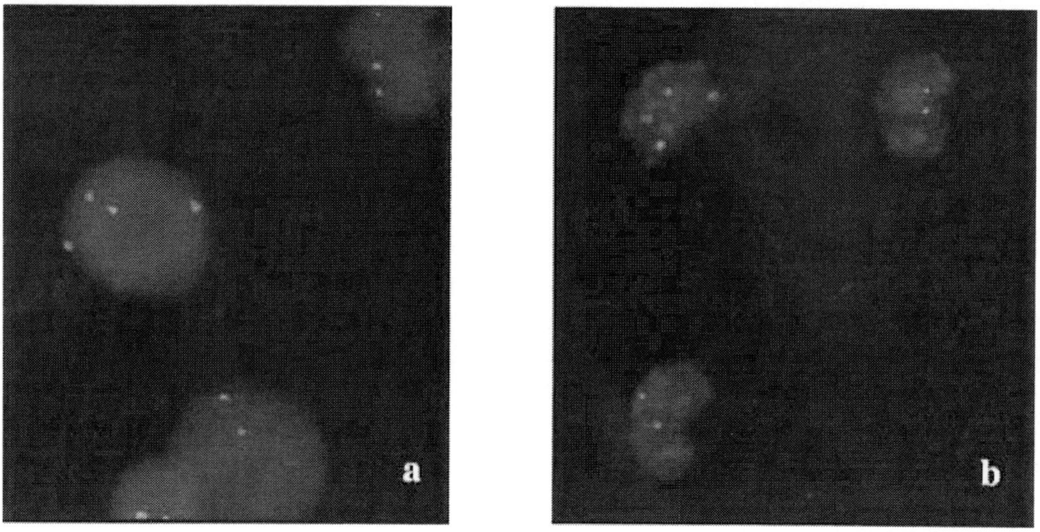

Figure 2a and 2b. (a) A male fetal nucleus displaying one green signal (Y-chromosome) and three red signals (chromosome18) surrounded by three maternal nuclei with two red signals. The analysis was performed with the dual-color PRINS reaction using a specific primer of the chromosome 18 centromere detected with an anti-digoxigenin antibody labeled to rhodamine and two specific primers for chromosome Y, detected using avidine-biotin antibody labeled to FITC. (b) Male fetal nucleus displaying one green signal (Y-chromosome) and three red signals (chromosome 21) surrounded by two maternal nuclei with two red signals. The technique was performed combining LSI 21 probe (SpectrumOrange) for FISH technique and two Y-specific primers detected using avidine-biotin antibody labeled to FITC for the PRINS reaction. Image capturing was performed at x1000 magnification

All false-positive signals were confirmed as false-positive signal using this approach. The assessment of chromosomes 13 and 21 by means of PRINS is difficult because the PRINS reaction for a single copy sequence is not available for these chromosomes. We had some success with PRINS reaction applied to a single-copy sequences but the technique is not sufficiently robust to be routinely used for FNCs detection. Furthermore, it has proven difficult to discriminate between sequence variations (polymorphisms) in the chromosome 13 and 21-satellite regions using PRINS [58]. We have developed a second protocol (Protocol 2) for the male trisomy 21 cells using a unique DNA sequence selected on chromosome 21 (FISH technique, probe LSI-21 SpectrumOrange) and sequences of the Y chromosome (PRINS reaction) (Figure 2a). Moreover, we used different approaches to confirm the fetal origin as specified before. All cases suspected to be false positive turned out to be false positive after FISH rehybridization. Thus, availability of specific primers for most of the commonly aneuploid chromosomes allows the PRINS technique to be used for aneuploid FNC identification.

Direct detection and analysis of intact FNCs from maternal blood could constitute one promising and exciting approach for a non-invasive alternative for prenatal diagnosis. So far, this approach has been hampered by the rarity of the FNCs (our results showed an average of 3 FNCs per mL of maternal blood in normal male pregnancies). The fact that the number of FNC detected in aneuploid pregnancies was 2 to 5-fold higher than in euploid pregnancies could be due to early impairment of the feto-maternal barrier.

PRINS and FISH detection methods have different advantages. We have shown that FISH and PRINS techniques had similar specificity and sensitivity when applied to interphase nuclei [59]. The PRINS reaction requires a better quality of sample's preparation than does FISH. Since the PRINS primers are not labeled, they can be used in excess without increasing background contrary to FISH. Furthermore, the PRINS technique is much faster than the FISH method; the entire PRINS procedure can be completed in one hour while the classical FISH procedure requires to 12 to 18 hours [46,57]. Due to the high cost of labeled FISH probes, materials for PRINS cost approximately ten times less than do materials for FISH [46]. Although it has been used for pre-implantation diagnosis [60], sperm analysis [61,62], and rapid prenatal diagnosis [48,63,64], the use of PRINS for the analysis of circulating fetal cells is still in its technological infancy [38,39,51,65]. In summary, compared to FISH, PRINS is more advantageous since it is a faster, easier and more cost-effective technique. At this stage of our investigations and comparatively to FISH technique, we noted a small number of requirements specific to PRINS reaction. Among them, PRINS reaction requires a better quality of the sample's preparation. Nonetheless, the potential exists for identifying fetal cells from maternal blood and then analyzing these cells for fetal genetic abnormalities.

Conclusion

The PRINS technique showed multiple advantages over FISH. While specificity and sensitivity were similar, PRINS was more rapid, easier, more reliable, and more cost-effectiveness than FISH. In addition, after PRINS, the slide can be subsequently reprobed with FISH to confirm or extend the PRINS result. PRINS also can be used to study single-copy genes and, can be routinely performed to investigate deletions associated with microdeletion syndromes. Therefore, the PRINS technique has the potential to become a widely used molecular cytogenetic tool in clinics and research. To isolate and analyze the extremely scarce fetal cells from maternal blood using, this very sensitive and specific alternative method is clearly one a valuable use for PRINS for an early cost-effective non-invasive prenatal cytogenetic test. The development of an automated system to detect rare PRINS signals would facilitate this goal. Although there are still many questions that remain unanswered, the outlook for the future use of PRINS for prenatal diagnosis is optimistic.

Acknowledgments

The authors are grateful to Dr. Gerald P. Holmquist for careful review of the manuscript and relevant comments. They also sincerely thank the women who made this study possible. This study was supported by a grant from the Canadian Institute for Health Research and the Canadian Genetic Diseases Network (MRC/NSERC NCE Program) to Régen Drouin. Dr. Kada Krabchi was a recipient of a studentship of "Valorisation Recherche Québec". Mr. Macoura Gadji was the recipient of a studentship from the "The Foundation for Research into Children's Diseases". Dr. Ju Yan is a research scholar (junior I level) of the « Fonds de la Recherche en Santé du Québec » program. Régen Drouin holds the Canada Research Chair in « Genetics, Mutagenesis and Cancer ».

References

[1] Schmorl G, Pathologisch-anatomische Untersurchungen über Publeraklampsie., ed. *Vogel.* 1893, Leipzig.

[2] Douglas GW, Thomas L, Carr M, Cullen NM, Morris R. Trophoblast in the circulating blood during pregnancy. *Am J Obstet Gynecol* 1959;78:960-73.

[3] Walknowska J, Conte FA, Grumbach MM. Practical and theoretical implications of fetal-maternal lymphocyte transfer. *Lancet* 1969;1:1119-22.

[4] Schroder J,De la Chapelle A. Fetal lymphocytes in the maternal blood. *Blood* 1972;39:153-62.

[5] Goldberg JD. Fetal cells in maternal circulation: progress in analysis of a rare event. *Am J Hum Genet* 1997;61:806-9.

[6] Bruch JF, Metezeau P, Garcia-Fonknechten N, Richard Y, Tricottet V, Hsi BL, Kitzis A, Julien C, Papiernik E. Trophoblast-like cells sorted from peripheral maternal blood using flow cytometry: a multiparametric study involving transmission electron microscopy and fetal DNA amplification. *Prenat Diagn* 1991;11:787-98.

[7] Cacheux V, Milesi-Fluet C, Tachdjian G, Druart L, Bruch JF, Hsi BL, Uzan S, Nessmann C. Detection of 47,XYY trophoblast fetal cells in maternal blood by fluorescence in situ hybridization after using immunomagnetic lymphocyte depletion and flow cytometry sorting. *Fetal Diagn Ther* 1992;7:190-4.

[8] Van Wijk IJ, Van Vugt JM, Konst AA, Mulders MA, Nieuwint AW, Oudejans CB. Multiparameter in situ analysis of trophoblast cells in mixed cell populations by combined DNA and RNA in situ hybridization. *J Histochem Cytochem* 1995;43:709-14.

[9] Adinolfi M. Non- or minimally invasive prenatal diagnostic tests on maternal blood samples or transcervical cells. *Prenat Diagn* 1995;15:889-96.

[10] Bianchi DW, Flint AF, Pizzimenti MF, Knoll JH, Latt SA. Isolation of fetal DNA from nucleated erythrocytes in maternal blood. *Proc Natl Acad Sci U S A* 1990;87:3279-83.

[11] Price JO, Elias S, Wachtel SS, Klinger K, Dockter M, Tharapel A, Shulman LP, Phillips OP, Meyers CM, Shook D, et al. Prenatal diagnosis with fetal cells isolated

from maternal blood by multiparameter flow cytometry. *Am J Obstet Gynecol* 1991;165:1731-7.

[12] Zheng YL, Carter NP, Price CM, Colman SM, Milton PJ, Hackett GA, Greaves MF, Ferguson-Smith MA. Prenatal diagnosis from maternal blood: simultaneous immunophenotyping and FISH of fetal nucleated erythrocytes isolated by negative magnetic cell sorting. *J Med Genet* 1993;30:1051-6.

[13] Ganshirt-Ahlert D, Burschyk M, Garritsen HS, Helmer L, Miny P, Horst J, Schneider HP, Holzgreve W. Magnetic cell sorting and the transferrin receptor as potential means of prenatal diagnosis from maternal blood. *Am J Obstet Gynecol* 1992;166:1350-5.

[14] Slunga-Tallberg A, el-Rifai W, Keinanen M, Ylinen K, Kurki T, Klinger K, Ylikorkala O, Knuutila S. Maternal origin of nucleated erythrocytes in peripheral venous blood of pregnant women. *Hum Genet* 1995;96:53-7.

[15] Slunga-Tallberg A,Knuutila S. Can nucleated erythrocytes found in maternal venous blood be used in the noninvasive prenatal diagnosis of fetal chromosome abnormalities? *Eur J Hum Genet* 1995;3:264-70.

[16] Troeger C, Zhong XY, Burgemeister R, Minderer S, Tercanli S, Holzgreve W, Hahn S. Approximately half of the erythroblasts in maternal blood are of fetal origin. *Mol Hum Reprod* 1999;5:1162-5.

[17] Zheng YL, Zhen DK, Farina A, Berry SM, Wapner RJ, Williams JM, Bianchi DW. Fetal cell identifiers: results of microscope slide-based immunocytochemical studies as a function of gestational age and abnormality. *Am J Obstet Gynecol* 1999;180:1234-9.

[18] Mesker WE, Ouwerkerk-van Velzen MC, Oosterwijk JC, Bernini LF, Golbus MS, Kanhai HH, Van Ommen GJ, Tanke HJ. Two-colour immunocytochemical staining of gamma (gamma) and epsilon (epsilon) type haemoglobin in fetal red cells. *Prenat Diagn* 1998;18:1131-7.

[19] Martin WL, Durrant LG, Liu DT. Non-invasive fetal cell isolation from maternal blood. *Br. J. Obstet Gynaecol* 1998;105:576-83.

[20] Pearson HA. Life-span of the fetal red blood cell. *J. Pediatr* 1967;70:166-71.

[21] Ganshirt D, Garritsen HS, Holzgreve W. Fetal cells in maternal blood. *Curr Opin Obstet Gynecol* 1995;7:103-8.

[22] Cha DH, Farina A, Bianchi DW, Johnson KL. ROC analysis of an erythroblast morphologic scoring system to improve identification of fetal cells in maternal blood. *Prenat Diagn* 2004;24:117-20.

[23] Hamada H, Arinami T, Sohda S, Hamaguchi H, Kubo T. Mid-trimester fetal sex determination from maternal peripheral blood by fluorescence in situ hybridization without enrichment of fetal cells. *Prenat Diagn* 1995;15:78-81.

[24] Hsieh TT, Pao CC, Hor JJ, Kao SM. Presence of fetal cells in maternal circulation after delivery. *Hum Genet* 1993;92:204-5.

[25] Schröder J, Tiilikainen A, De la Chapelle A. Fetal leukocytes in the maternal circulation after delivery. *Transplantation* 1994;17:346-54.

[26] Bianchi DW, Zickwolf GK, Weil GJ, Sylvester S, DeMaria MA. Male fetal progenitor cells persist in maternal blood for as long as 27 years postpartum. *Proc Natl Acad Sci USA* 1996;93:705-8.

[27] Jacobs PA,Smith PG. Practical and theoretical implications of fetal-maternal lymphocyte transfer. *Lancet* 1969;2:745.

[28] 28. Polani PE,Mutton DE. Y-fluorescence of interphase nuclei, especially circulating lymphocytes. *Br Med J* 1971;1:138-42.

[29] 29. Bianchi DW, Williams JM, Sullivan LM, Hanson FW, Klinger KW, Shuber AP. PCR quantitation of fetal cells in maternal blood in normal and aneuploid pregnancies. *Am J Hum Genet* 1997;61:822-9.

[30] 30. Evans PC, Lambert N, Maloney S, Furst DE, Moore JM, Nelson JL. Long-term fetal microchimerism in peripheral blood mononuclear cell subsets in healthy women and women with scleroderma. *Blood* 1999;93:2033-7.

[31] Artlett CM, Smith JB, Jimenez SA. Identification of fetal DNA and cells in skin lesions from women with systemic sclerosis. *N Engl J Med* 1998;338:1186-91.

[32] Nelson JL, Furst DE, Maloney S, Gooley T, Evans PC, Smith A, Bean MA, Ober C, Bianchi DW. Microchimerism and HLA-compatible relationships of pregnancy in scleroderma. *Lancet* 1998;351:559-62.

[33] Zilliacus R, De la Chapelle A, Schroder J, Tilikainen A, Kohne E, Kleihauer E. Transplacental passage of foetal blood cells. *Scand J Haematol* 1975;15:333-8.

[34] Wessman M, Ylinen K, Knuutila S. Fetal granulocytes in maternal venous blood detected by in situ hybridization. *Prenat Diagn* 1992;12:993-1000.

[35] Bianchi DW. Prenatal diagnosis by analysis of fetal cells in maternal blood. *J Pediatr* 1995;127:847-56.

[36] Bianchi DW, Mahr A, Zickwolf GK, Houseal TW, Flint AF, Klinger KW. Detection of fetal cells with 47,XY,+21 karyotype in maternal peripheral blood. *Hum Genet* 1992;90:368-70.

[37] Ganshirt-Ahlert D, Borjesson-Stoll R, Burschyk M, Dohr A, Garritsen HS, Helmer E, Miny P, Velasco M, Walde C, Patterson D, et al. Detection of fetal trisomies 21 and 18 from maternal blood using triple gradient and magnetic cell sorting. *Am J Reprod Immunol* 1993;30:194-201.

[38] Krabchi K, Gros-Louis F, Yan J, Bronsard M, Masse J, Forest JC, Drouin R. Quantification of all fetal nucleated cells in maternal blood between the 18th and 22nd weeks of pregnancy using molecular cytogenetic techniques. *Clin Genet* 2001;60:145-50.

[39] Orsetti B, Lefort G, Boulot P, Andreo B, Pellestor F. Fetal cells in maternal blood: the use of primed in situ (PRINS) labelling technique for fetal cell detection and sex assessment. *Prenat Diagn* 1998;18:1014-22.

[40] Krabchi K, Gadji M, Yan J, Drouin R, Dual-Color PRINS for In Situ Detection of Fetal Cells in Maternal Blood, in *PRINS and In Situ PCR Protocols*, F Pellestor, Editor. 2006, Humana Press: Totowa, NJ. p. 141-9.

[41] Camaschella C, Alfarano A, Gottardi E, Travi M, Primignani P, Caligaris Cappio F, Saglio G. Prenatal diagnosis of fetal hemoglobin Lepore-Boston disease on maternal peripheral blood. *Blood* 1990;75:2102-6.

[42] Kao SM, Tang GC, Hsieh TT, Young KC, Wang HC, Pao CC. Analysis of peripheral blood of pregnant women for the presence of fetal Y chromosome-specific ZFY gene deoxyribonucleic acid sequences. *Am J Obstet Gynecol* 1992;166:1013-9.

[43] Lo YM, Patel P, Baigent CN, Gillmer MD, Chamberlain P, Travi M, Sampietro M, Wainscoat JS, Fleming KA. Prenatal sex determination from maternal peripheral blood using the polymerase chain reaction. *Hum Genet* 1993;90:483-8.

[44] Koch JE, Kolvraa S, Petersen KB, Gregersen N, Bolund L. Oligonucleotide-priming methods for the chromosome-specific labelling of alpha satellite DNA in situ. *Chromosoma* 1989;98:259-65.

[45] Koch J, Hindkjaer J, Kolvraa S, Bolund L. Construction of a panel of chromosome-specific oligonucleotide probes (PRINS-primers) useful for the identification of individual human chromosomes in situ. *Cytogenet Cell Genet* 1995;71:142-7.

[46] Yan J, Bronsard M, Drouin R. Creating a new color by omission of 3 end blocking step for simultaneous detection of different chromosomes in multi-PRINS technique. *Chromosoma* 2001;109:565-70.

[47] Yan J, Gadji M, Krabchi K, Drouin R, New rapid Multicolor PRINS Protocol, in *PRINS and In Situ PCR Protocols*, F Pellestor, Editor. 2006, Humana Press: Totowa, NJ. p. 3-13.

[48] Pellestor F, Girardet A, Lefort G, Andreo B, Charlieu JP. Use of the primed in situ labelling (PRINS) technique for a rapid detection of chromosomes 13, 16, 18, 21, X and Y. *Hum Genet* 1995;95:12-7.

[49] Speel EJ, Lawson D, Hopman AH, Gosden J. Multi-PRINS: multiple sequential oligonucleotide primed in situ DNA synthesis reactions label specific chromosomes and produce bands. *Hum Genet* 1995;95:29-33.

[50] Krabchi K, Gadji M, Samassekou O, Gregoire MC, Forest JC, Drouin R. Quantification of fetal nucleated cells in maternal blood of pregnant women with a male trisomy 21 fetus using molecular cytogenetic techniques. *Prenat Diagn* 2006;26:28-34.

[51] Krabchi K, Gadji M, Forest JC, Drouin R. Quantification of all fetal nucleated cells in maternal blood in different cases of aneuploidies. *Clin Genet* 2006;69:145-54.

[52] Lewis DE, Schober W, Murrell S, Nguyen D, Scott J, Boinoff J, Simpson JL, Bischoff FZ, Elias S. Rare event selection of fetal nucleated erythrocytes in maternal blood by flow cytometry. *Cytometry* 1996;23:218-27.

[53] de Graaf IM, Jakobs ME, Leschot NJ, Ravkin I, Goldbard S, Hoovers JM. Enrichment, identification and analysis of fetal cells from maternal blood: evaluation of a prenatal diagnosis system. *Prenat Diagn* 1999;19:648-52.

[54] de Graaf IM, van Bezouw SM, Jakobs ME, Leschot NJ, Zondervan HA, Bilardo CM, Hoovers JM. First-trimester non-invasive prenatal diagnosis of triploidy. *Prenat Diagn* 1999;19:175-7.

[55] Sekizawa A, Samura O, Zhen DK, Falco V, Farina A, Bianchi DW. Apoptosis in fetal nucleated erythrocytes circulating in maternal blood. *Prenat Diagn* 2000;20:886-9.

[56] van Wijk IJ, de Hoon AC, Jurhawan R, Tjoa ML, Griffioen S, Mulders MA, van Vugt JM, Oudejans CB. Detection of apoptotic fetal cells in plasma of pregnant women. *Clin Chem* 2000;46:729-31.

[57] Hindkjaer J, Brandt CA, Stromkjaer H, Koch J, Kolvraa S, Bolund L. Primed IN situ labelling (PRINS) as a rational procedure for identification of marker chromosomes using a panel of primers differentially tagging the human chromosomes. *Clin Genet* 1996;50:437-41.

[58] Pellestor F, Girardet A, Lefort G, Andreo B, Charlieu JP. Selection of chromosome-specific primers and their use in simple and double PRINS techniques for rapid in situ identification of human chromosomes. *Cytogenet Cell Genet* 1995;70:138-42.

[59] Yan J, Zhang XX, Fetni R, Drouin R. Trisomy 8 and monosomy 7 detected in bone marrow using primed in situ labeling, fluorescence in situ hybridization, and conventional cytogenetic analyses. A study of 54 cases with hematological disorders. *Cancer Genet Cytogenet* 2001;125:30-40.

[60] Pellestor F, Girardet A, Andreo B, Lefort G, Charlieu JP. Preimplantation embryo chromosome analysis by primed in situ labeling method. *Fertil Steril* 1996;66:781-6.

[61] Coignet L, Girardet A, Andreo B, Charlieu JP, Pellestor F. Double and triple in situ chromosomal labeling of human spermatozoa by PRINS. *Cytogenet Cell Genet* 1996;73:300-3.

[62] Pellestor F, Girardet A, Coignet L, Andreo B, Charlieu JP. Assessment of aneuploidy for chromosomes 8, 9, 13, 16, and 21 in human sperm by using primed in situ labeling technique. *Am J Hum Genet* 1996;58:797-802.

[63] Velagelati GV, Shulman LP, Phillips OP, Tharapel SA, Tharapel AT. Primed in situ labeling for rapid prenatal diagnosis. *Am J Obstet Gynecol* 1998;178:1313-20.

[64] Gadji M, Krabchi K, Drouin R. Simultaneous identification of chromosomes 18, X and Y in uncultured amniocytes by using multi-primed in situ labelling technique. *Clin Genet* 2005;68:15-22.

[65] Krabchi K, Lavoie J, Coullin P, Bronsard M, Pellestor F, Yan J, Drouin R. [From the conception of the PRINS to its coronation]. *Med Sci* (Paris) 2004;20:465-73.

In: PRINS and PNA Technologies in Chromosomal Investigations
Editor: Franck Pellestor, pp. 53-67
ISBN 1-60021-413-4

Chapter 5

PRINS Detection of Centromeric and Telomeric Sequences in Micronuclei: A Useful Tool to Distinguish Chromosome Breakage from Aneuploidy

Antonella Russo *

Department of Biology, University of Padova, Padova, Italy

Abstract

Micronuclei are indirect but reliable indicators of chromosome damage, formed at the mitotic division after either chromosome loss or chromosome breakage events, and visible in the cytoplasm of interphase cells as small bodies of extra-chromatin. Centromeric, pericentromeric, or telomeric repetitive sequences can be considered as indicators of chromosome integrity into the micronucleus. Therefore the localisation of these sequences may contribute to predict the mechanism of induction of micronuclei following the exposure of mammalian cells to mutagens. We review here the results obtained in our laboratory by application of different PRINS (PRimed In Situ labelling) strategies for characterising the micronucleus content; the performance of PRINS is compared with that of the other procedures (kinetochore immunodetection, Fluorescence In Situ Hybridisation) applicable to the same goal.

Keywords: Micronucleus, PRINS, chromosome aberrations, aneuploidy, chromosome breakage, environmental mutagenesis

* Correspondence to: Dr. Antonella Russo, Department of Biology, Via U. Bassi 58/b, 35131 Padova, Italy. Phone: +39 049 827 6279. Fax: +39 049 827 6280. E-mail: antonella.russo@unipd.it

Introduction

Many physical and chemical agents present in our environment are able to induce DNA damage and to increase chromosome aberration and mutation frequencies. Environmental mutagenesis is the field aimed to define the way of action of potential mutagenic agents, by means of a panel of assays detecting different genomic lesions, included structural and numerical aberrations. Dose-effect relationships may be defined for better understanding the biological activity of the compounds of interest. Either structural or numerical chromosome aberrations can be induced after exposure to several mutagens; although the pathways leading to these abnormalities (as well as the primary lesions involved) are different, a number of physical and chemical agents of environmental interest were demonstrated able to induce both types of chromosome damage. A classical example is represented by ionising radiation (IR), which induces very efficiently the primary DNA lesion leading to chromosome break: the double strand breaks (DSB). For that reason IR is among the most powerful clastogenic (i.e. affecting the chromosome structure) agents. IR can however injure directly or indirectly other cell targets, which are crucial for genome stability; by these pathways, IR can induce also numerical aberrations [1-3].

Direct chromosome analysis is laborious, time-consuming and requires skilled personnel for being conducted. With the aim to develop an experimental methodology providing good sensitivity, but assuring a fast accumulation of data concerning the induction of chromosome damage, the micronucleus (MN) assay has been proposed about 30 years ago [4-5]. Micronuclei are small chromatin cytoplasmatic bodies, which can arise spontaneously in dividing cells bearing a chromosome fragment, or undergoing a segregation error (Fig. 1). Although MN are merely indirect indicators of chromosome damage, it has been demonstrated that their frequency increases in cell populations exposed either *in vivo* or *in vitro* to different radiations or chemicals. Micronuclei are formed at the mitotic (or meiotic) division by two different mechanisms depicted in Fig. 1: if an acentric chromosome fragment is present at the moment of the cell division, it will be possibly excluded from both daughter nuclei, remaining visible as a MN in the next interphase (Fig. 1, a); in alternative, micronuclei can derive from chromosome lagging at anaphase, as described in Fig. 1, b. In its original proposition, the MN assay was carried out on mouse bone marrow young (polychromatic) erythrocytes, and it provided a simple *in vivo* experimental procedure for the identification of chemical and physical agents affecting genome stability [4-5]. The major advantage of the MN assay in mouse polychromatic erythrocytes was represented by the fact that the MN detection is immediate, because in this developmental cell stage the main nucleus is lacking. Therefore the assay does not require skilled personnel, in contrast to the direct chromosome aberration analysis in metaphase spreads.

Several strategies were then proposed to apply the same rationale to other cell types growing *in vivo* or *in vitro*. A crucial requirement for the conduction of a MN assay is that the cell line to be investigated must be proliferating. The approximate length of each phase of the cell cycle should be also known: in this way, it is possible to define an appropriate time interval between cell exposure and cell harvesting, taking into account that MN arise at mitosis, but the damage is induced in a previous cell cycle phase (according to the mechanism of action of the agent under investigation).

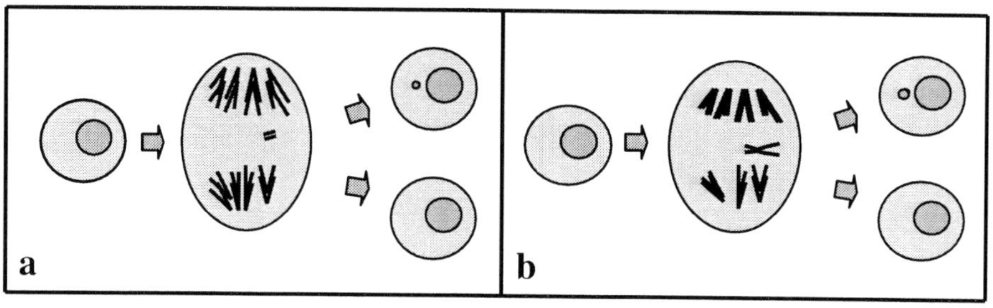

Figure 1. The two pathways leading to micronucleus formation after a mitotic cell division. (a) A cell carrying a chromosome break cannot control the segregation of the acentric fragment, which can be excluded from both daughter nuclei. In the cell progeny, one cell will carry a micronucleus. (b) A chromosome segregation error can occur during the mitotic cell division; if a chromosome is not able to keep/maintain contact with the spindle, it will lag at anaphase and possibly will be excluded from both daughter nuclei. In the cell progeny, one cell will carry a micronucleus. Note that the two mechanisms lead to the same visible result

Moreover, for an accurate evaluation of MN frequencies it would be important to discard from the analysis those cells which did not pass through mitosis after the treatment, because an eventual damage cannot be already expressed as a MN in these cells. This prerequisite is more easily accomplished in an *in vivo* test, where appropriate developmental cell stages can be chosen for the analysis (for example, young erythrocytes as the cells deriving from the last differentiating mitotic division of erythroblasts, early spermatids as the cells deriving from the two subsequent meiotic divisions). For *in vitro* analysis, the use of cytochalasin B, an inhibitor of cell cytokinesis which induces the formation of binucleated cells has been proposed, in order to restrict the analysis to the representative cell sample of cells (the binucleated cells) [6-7]. However, a MN will appear as a round chromatin body in interphase cells whichever the mechanism (clastogenic or aneuploidogenic one) leading to its formation; because the MN deriving from a chromosome segregation error should contain the whole chromosome, the presence of the centromere into the MN can be assessed to predict the mechanism(s) of action of the agent under study (Fig. 2, left panel). Indeed, the MN assay was combined with the molecular detection of the centromeric region as soon as the appropriate approaches (kinetochore immunodetection, Fluorescence In Situ Hybridisation (FISH) with centromeric probes) became available to the scientific community in the middle of ‘80s [8-14]. Advantages and disadvantages of these approaches are discussed in [15].

An alternative indicator of the integrity of the chromosome is the number of telomeres (Fig. 2, right panel); FISH detection of telomeric sequences into observed MN has been applied however only sporadically to distinguish aneploidogenic from clastogenic action of environmental mutagens [16-17].

The molecular organisation of the repetitive sequences mapping at the centromeric or telomeric regions deeply vary among species and cell types: in humans, the centromeric alpha satellite DNA is constituted by a 171 bp repeat, which is largely polymorphic among chromosomes, a feature allowing the detection of single chromosome pairs by means of the appropriate probe [12]. A consensus sequence of 17-bp known as the CENP-B box has been described [18].

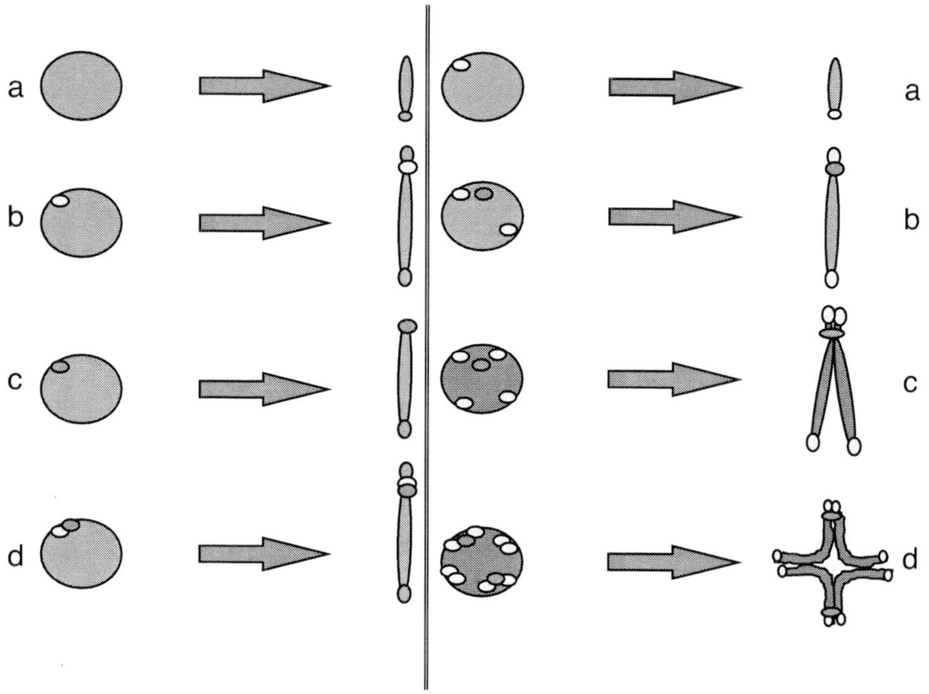

Figure 2. The detection of repetitive sequences into micronuclei allows making inference on the mechanism of origin. On the left panel, the detection of centromeric sequences in a single colour approach allows to distinguish micronuclei representing acentric fragments (centromere-negative MN, a), from micronuclei harbouring a whole chromosome and due to chromosome loss (centromere-positive MN, b). If a tandem labelling approach with mouse centromeric and pericentromeric sequences is considered, three classes of micronuclei are expected (a, c, d): acentric fragments will be contained either in unlabelled MN (a) or in MN carrying the pericentromeric satellite sequence only, and representing huge chromosome fragments formed by breakage into the pericentromeric heterochromatin (c); MN positive for both sequences will represent whole chromosomes derived from a segregation error (d). On the right panel, the application of dual colour detection of centromeric and telomeric sequences is summarised. Micronuclei deriving from chromosome breakage will not show the centromeric sequence but one or two telomeric sequences should be observed, as expected for acentric chromatid (a) or chromosome (not shown) fragments. Centromere-positive MN are formed as the consequence of a segregation error; depending if the event has occurred before or during anaphase the number of telomeres observed in these MN can be 2-4 (b, c). In (d) the special case of a MN scored in postmeiotic cells and derived from the loss of a whole bivalent is shown

This sequence can be used as target for a pancentromeric probe [13, 14]. Pericentromeric regions of human chromosomes also show a very high degree of variation: several classes of repetitive DNA are described, and the chromosome-specific organization is known [19]. In contrast, in the mouse a single class of repetitive DNA, the so called minor satellite DNA, can be detected at the centromeres of all autosomes and the X chromosome, while a larger repetitive array (for that reason dubbed major satellite DNA] is located at the pericentromeric region of the same chromosomes [20]. The minor DNA satellite repetitive sequence shows no homology with the alphoid repeat, but notably it contains the CENP-B box [18]. The major DNA satellite represents 5-10% of the whole mouse genome [20]. All vertebrates share a conserved telomeric sequence, but variable length of the repetitive array among chromosomes, cell types, species is observed.

As data were growing concerning the characterisation of induced MN in mammalian cells, it appeared that most agents can induce both chromosome loss and breakage [15]. In addition, after the introduction of multicolour FISH, the pericentromeric heterochomatin was found to be a preferential target for some clastogenic agents [21-23]. The tandem labelling approach [24-25] consists in the dual-colour localisation of the centromeric sequence and its adjacent pericentromeric repetitive DNA (Fig. 2, left panel). MN containing a whole chromosome should appear as double labelled (Fig. 2, left panel, d), while those deriving from an acentric chromosome fragment should be unlabelled (a), unless the break have occurred into the repetitive sequence located at the pericentromeric region; only in this case the MN will show a single colour labelling corresponding to the retained portion of the pericentromeric satellite DNA (c). This approach is of great utility for mouse cells, while for human cells, where a pancentromeric probe is available but the pericentromeric region of different chromosomes are highly heterogeneous in sequence and organization, the tandem labelling approach is applied with a different experimental strategy with respect to the MN assay [25].

In our laboratory, we were interested in the possibility to achieve a fast detection of centromeric sequences as well as telomeric ones by PRINS (PRimed IN Situ labelling), and to apply this approach to the micronucleus assay in several human and murine cell types.

There were different reasons for the choice of PRINS as alternative to FISH: the main one was the opportunity to develop a protocol easily reproducible on cell types of different species. This aspect is of great value, since the optimisation of commercially available probes for FISH is mainly restricted to diagnostic purposes, and therefore to the human genome. Instead, great interest for model species exists in the field of environmental mutagenesis. Second, PRINS is faster than FISH, the experimental procedure is completed in a single day, and the success of the procedure can be checked immediately. Third, the background noise is lower with PRINS than FISH, which is crucial in interphase analysis when the discrimination of true reaction signals from aspecific fluorescent spots is difficult and can lead to wrong conclusions.

Methods

Several cell types were considered to combine the micronucleus assay with PRINS detection of DNA repetitive sequences.

The following cell types were isolated from the mouse after *in vivo* treatment: bone marrow erythrocytes, splenocytes, germ cells. Bone marrow has been processed and analysed as described in [26]. The micronucleus assay in germ cells has been conducted in early spermatids, to evaluate the effects of a treatment in meiosis [27]. Splenocytes are quiescent cells which can stimulated to enter the cell cycle in the presence of concanavalin A (ConA), to study the consequences of an *in vivo* exposure of the resting cells, or to assay the effects of *in vitro* treatments before and after ConA stimulation [23]; in the latter case, splenocytes are obtained from untreated mice.

Several murine and human cell lines were used for *in vitro* assays. Either primary or immortalised cells were used; they were treated and harvested following the guidelines of the *in vitro* micronucleus assay [7].

Slides were obtained by cytocentrifugation; the preparations were then fixed in cold ethanol, and stored in fresh absolute ethanol at -20° C (more details in: [28]). Single-colour or dual-colour detection of DNA repetitive sequences was applied to achieve the following goals:

a) to define the clastogenic and/or aneuploidogenic potential of each suspected mutagen. A single-colour detection of centromeric sequences was applied for this specific purpose (Fig. 2, left panel, a-b). MN showing positive labelling for the centromeric sequence (C+ MN), and MN appearing negatively labelled (C- MN) are enumerated on an adequate cell sample (at least 1000 cells per experimental condition and per replicated experiment). If the treatment is inducing chromosome breaks, the frequency of C- MN in treated cell samples will result significantly higher with respect to the control; an increase of C+ MN will indicate that the treatment is inducing chromosome segregation errors. Notably, the two effects are not mutually exclusive, as many agents can give either structural or numerical aberrations; the major effect (clastogenesis or aneuploidy) induced by the agent will be evaluated by calculating the C+ and C- MN relative proportions (agents acting with a prevalent clastogenic mechanism will produce a large majority of centromere-negative micronuclei; on the contrary, aneuploidogenic agents are expected to increase the proportion of centromere-positive micronuclei).
b) to verify if the pericentromeric heterochromatin is a preferential target for DNA breakage under specific treatment conditions. In this case, a dual colour approach aimed to simultaneously detect centromeric and pericentromeric DNA repetitive sequences of the mouse genome (namely minor satellite DNA and major satellite DNA) was applied (Fig. 2, left panel, a, c-d). Chromosome segregation errors should give origin to MN harbouring the whole chromosome, therefore carrying both classes of repetitive sequences (C++ MN, two adjacent fluorescent signals, with different colour). MN showing the pericentromeric major satellite DNA only (Cm MN), or resulting unlabelled after PRINS reaction (C- MN), will be considered both acentric fragments, however the former category will be specifically originated by breakage events induced in the pericentromeric heterochomatin. If these events are randomly occurring, the proportion of Cm MN on the total of MN arisen from DNA breaks (Cm MN plus C- MN) should correspond to the percentage of representation of the major satellite DNA on the whole genome (about 10%) [20]. Higher than expected Cm MN proportions are suggestive for preferential clastogenic action of the tested agent at the pericentromeric heterochromatin;
c) to verify the preferential mechanism by which a suspected aneuploidogenic agent can affect the chromosome segregation process, the simultaneous detection of centromeric and telomeric sequences was applied (Fig. 2, right panel). MN failing to incorporate in the two daughter nuclei may represent the consequence of different kind of damages affecting one of more of the cell targets which are involved in the

mitotic (or meiotic) process. If chromosomes are unable to align at the metaphase plate to be connected with the bipolar spindle, they could form a MN in the absence of chromatid splitting. In this case the MN will contain 4 telomeres together with the centromeric sequence (Fig. 2, right panel, c). Under the more conventional mechanism of chromosome loss due to anaphase lagging, the centromere-positive MN will contain only 2 telomeres (Fig. 2, right panel, b). The application of this experimental strategy to the germ cells can be complicated by the hypothesis of the loss of a whole bivalent into the MN (this lead to the observation of MN with up to 8 fluorescent telomeric spots; Fig. 2, right panel, d). MN must be labelled with at least one telomeric spot, in the hypothesis that they contain an acentric chromatid fragment (Fig. 2, right panel, a).

All the protocols are described in detail in [28].

Fluorescence in situ hybridisation (FISH) and kinetochore immunodetection (CREST) were applied following standard protocols; more details in [23].

Results

Model mutagens were used to evaluate the performance of PRINS, FISH and CREST approaches in *in vivo* and *in vitro* cell systems, and to make comparisons among those approaches.

Primary cultures of mouse splenocytes were set up for the application of the MN test in combination with FISH [23] or PRINS [29], after *in vitro* exposure to mitomycin C. The aim of these studies was to clarify the significance of previous observations concerning the effects of mitomycin C, a well known and powerful clastogenic agent, which in the mouse bone marrow assay combined with kinetochore immunodetection (CREST) appeared able to induce, besides the expected centromere-negative MN, a certain percentage of centromere-positive MN [30]. On the other hand, evidence was being collected in the same period that heterochomatic regions of the genome could be preferential targets for some clastogenic compounds [21-23]. Therefore we were particularly interested in the development of a tandem labelling PRINS approach [26]. By comparing the results obtained with the two molecular approaches [23, 29], it appeared that the class of MN bearing the pericentromeric (major satellite DNA) sequence, but not the centromeric one (minor satellite DNA), resulted underestimated with FISH with respect to PRINS [29]. The different outcome of the two approaches was reasonably the consequence of a less efficient detection of the major satellite DNA after hybridisation than primer extension, most probably on those MN retaining only few sequence arrays of the repetitive sequence [29]. On the other hand, C++ MN frequencies obtained after FISH and PRINS appeared highly comparable [23, 29]. Therefore the developed PRINS protocol resulted to be accurate; since higher sensitivity was obtained for PRINS than FISH, for our subsequent studies we used routinely PRINS instead of FISH.

Then, we have attempted the application of the PRINS tandem labelling approach in combination to the well established mouse bone marrow MN test [26], and to the MN assay in mouse early spermatids [27]. As anticipated in Introduction, the main advantage of the bone marrow MN test is that the analysis is conducted in young anucleated erythrocytes, which makes very easy the MN detection. In contrast, the molecular characterization of MN in the same cell type has some disadvantages; in particular, in the lack of the main MN an internal reference for the labelling quality is missing, therefore scoring of areas with faint fluorescent spots can lead to false negatives, while the presence of high background can lead to false positives. Another crucial factor is that the denaturation step can alter the MN morphology, leading to a difficult detection of fluorescent spots. Indeed, our unpublished data suggest a poor reproducibility with this kind of approach. Highly valuable were on the other hand the results obtained by PRINS tandem labelling in mouse early spermatids: the quality of the fluorescent spots was very good, and it allowed an easy classification of MN (Fig. 3). An example of the application of this approach concerned the effects of 4-aminobiphenyl (4-AB) on mouse germ cells. We tested this compound in the frame of a much wide study, and we observed that treatments on different meiotic stages produced a doubling of the baseline MN frequency. Unfortunately, in this cell system weak increases of MN with respect to the control are usually detected, a feature which makes it difficult to reach statistical confirmation for true effects. Even for the experiments described in Fig. 4, the sample size analysed was large (at least 16000 spermatids per experimental condition, obtained from three different mice) but it should be further improved to get conclusive results concerning the meiotic effects of 4-AB. In this situation, application of the PRINS methodology was helpful for a preliminary evaluation, as the analysis of the molecular content indicates that Cm MN and C- MN were elevated in spermatids from treated animals with respect to the control (Fig. 4). Therefore 4-AB may exert a weak clastogenic activity on different meiotic stages. Notably, Cm MN deriving form chromosome breakage into the pericentromeric heterochomatin were detected only in treated samples (Fig. 4).

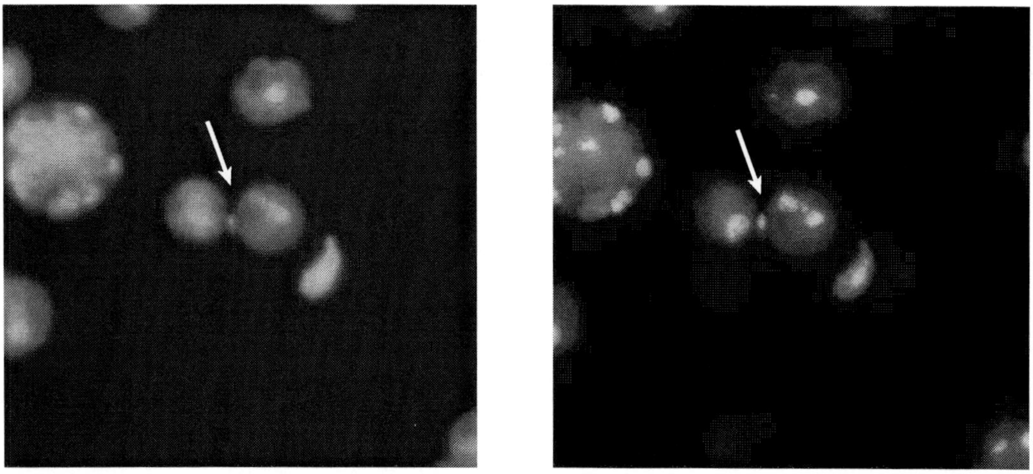

Figure 3. Tandem labelling of centromeric and pericentromeric sequences in a mouse early spermatid carrying a MN (arrow). Left: DAPI staining visualized under the UV filter. Right: triple band pass filter vision of the same cells; the centromeric (minor satellite DNA) sequence is labelled in yellow fluorescence, the pericentromeric (major satellite DNA) sequence is labelled in red fluorescence. As the MN contains both fluorescent signals, its formation can be ascribed to a chromosome segregation error occurred during the meiotic divisions

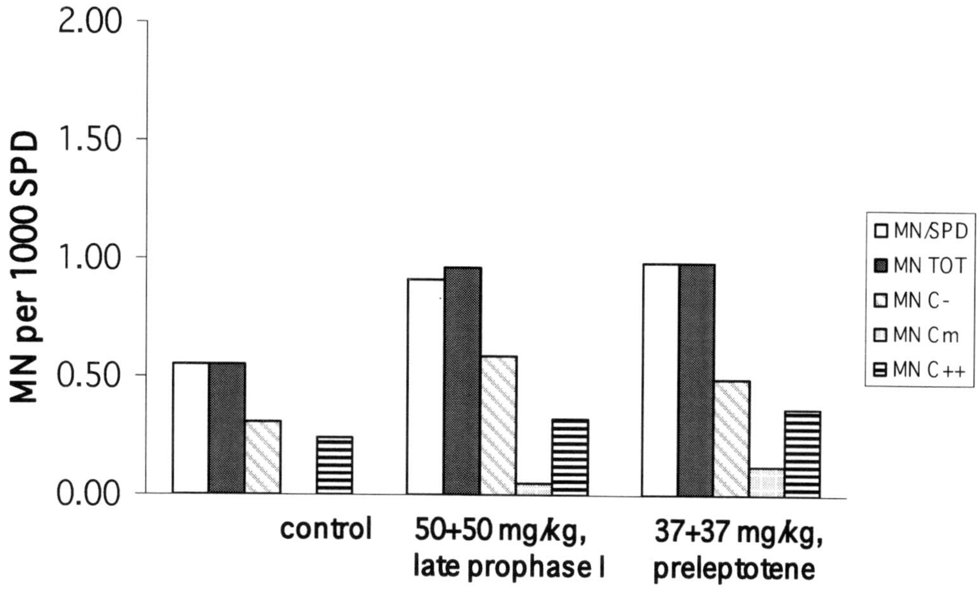

Figure 4. The effects of 4-aminobiphenyl on meiotic cells of the mouse, as evaluated by tandem labelling in early spermatids. Repeated treatments (24 h of interval) were assayed on different meiotic stages. At least 16000 early spermatids from 3 different mice were analysed per experimental point. It appears that after treatment of both meiotic stages, the total MN frequency is doubled with respect to the controls; however a statistical conclusion cannot be reached due to low MN frequencies detected in the cell sample considered. It can be noticed however that MN deriving from chromosome break (C-MN and Cm MN) are the MN categories increased in treated samples with respect to controls, thus suggesting that the weak effect observed is due to clastogenic action of the tested compound

We have then carried out a single colour analysis in human cells, to compare PRINS with the widely applied CREST approach, consisting in the kinetochore immunodetection. For that purpose, we took advantage from the preparations obtained during a much complex study on the combined effects of ionising radiation and microgravity conditions [31]; PRINS and CREST analysis was carried out in parallel on preparations obtained from the same cell samples. Centromeres were labelled by means of a primer designed on the 17-bp motif of the CENP-B box [28]. An example of the PRINS labelling of this sequence is given in Figure 5, together with an example of CREST staining of kinetochores. C+ and C- MN frequencies were evaluated 24 h after the irradiation of TK6 cells, a human lymphoblastoid cell line with nearly diploid chromosome content. Figure 6 shows the results obtained in TK6 control cultures and after irradiation with 2 Gy gamma rays; the data refer to three replicated experiments and to the analysis of 1000 cells per replica. Both PRINS and FISH methodologies were able to detect increases of C-MN as well as of C+ MN.

Finally, on the basis of successful application of dual colour PRINS approach, we attempted to develop a protocol for the detection of centromeric and telomeric sequences. In this case four categories of micronuclei are expected: centromere-positive MN can be due segregation errors before or during anaphase, and in the former situation they should harbour 4 telomeres, while in the latter one only 2 (Fig. 2, right panel b-c); the centromere negative MN should represent acentric chromosome fragments if carrying 2 telomere spots, or acentric chromatid fragments if carrying one telomeric spot only (Fig. 2, right panel a). We have provided evidence for the reproducibility of this approach by carrying out the characterisation of MN in the same mouse splenocyte preparations previously used for tandem labelling analysis [29, 32]. This parallel study gave us the possibility to verify that the frequencies of centromere-positive and centromere-negative MN were strictly comparable after being evaluated by different detection schemes: in particular, when tandem labelling is applied the centromeric sequence is elongated during the first PRINS reaction, while the same sequence is elongated during the second reaction when detected together with the telomeres. Therefore, it can be concluded that the dual colour PRINS reaction does not lead to underestimation of the second labelled sequence, even in the case of the relatively small signals of the minor satellite DNA. On the other hand, the results obtained in mouse splenocytes and subsequently on other cell types [33, 34] represented a direct confirmation that the first reaction is efficiently stopped before the second is started, as all observed MN showed (if any) a unique centromeric spot, distinct from the telomeric ones.

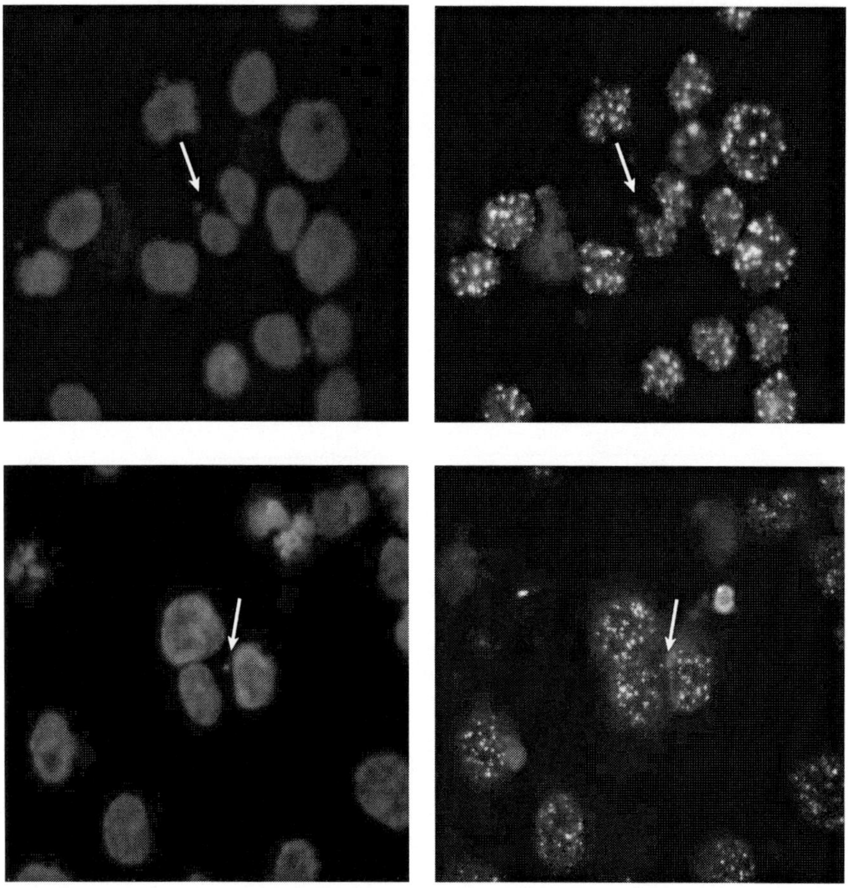

Figure 5. Centromere detection (yellow fluorescent spots) in micronucleated TK6 cells exposed to gamma irradiation. Top, left: a cell carrying a MN is indicated by the white arrow; top, right: PRINS centromere detection by CENP-B box labelling indicates that the MN contains a whole chromosome. Bottom, left: a cell carrying a MN is indicated by the white arrow; bottom, right: the MN appears to be positive to the CREST immunodetection of kinetochores, and therefore is harbouring a whole chromosome

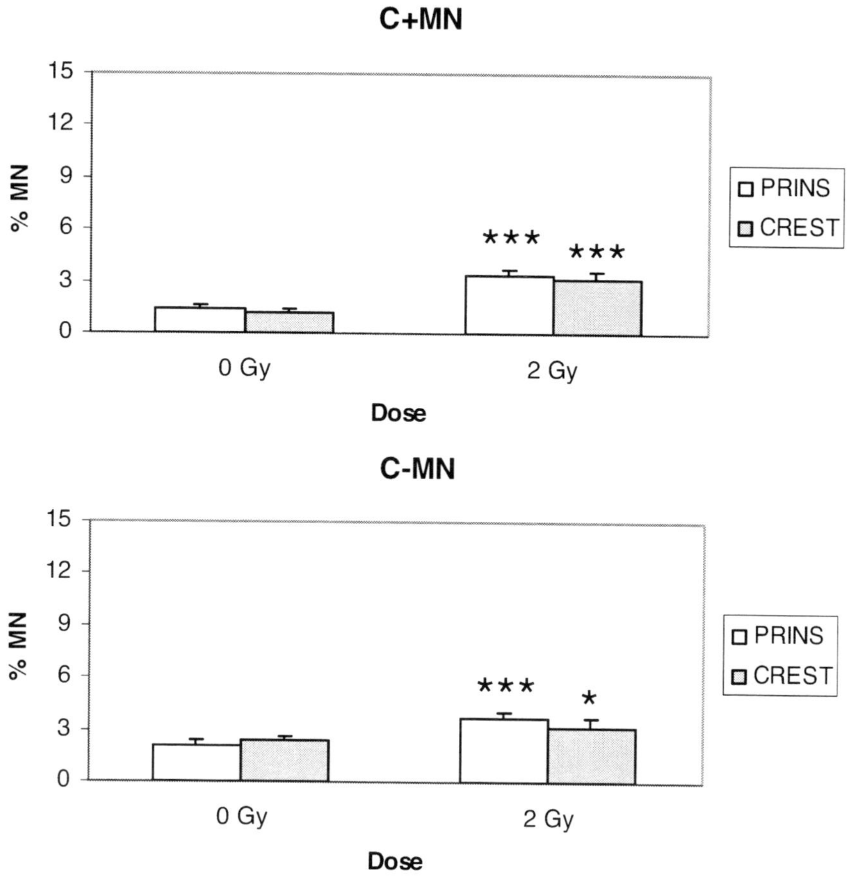

Figure 6. A comparison of the performance of PRINS and CREST methodologies for the evaluation of gamma irradiation effects in the human cell line TK6. * = $P < 0.05$; *** = $P < 0.001$

More importantly, all the results obtained with this approach [32-34] were consistent with the predicted mechanism of action of the agents under study, i.e. clastogenic agents specifically increased the frequency of centromere-negative MN (which as expected were prevalently showing one telomeric spot only), and aneuplodogenic agents produced an increase of centromere-positive MN, all of them carrying at least 2 telomeres in accordance to the prediction of the assay (Fig. 2, right panel). Remarkably, all observed MN in the diverse cell types analysed, carried at least one telomeric signal, underlying the good performance of the PRINS labelling [32, 34]. In contrast, the sensitivity of FISH in telomere detection into MN cells was reported to be poor [16-17].

The extension of the dual labelling approach of centromeres and telomeres in application of the spermatid MN assay [34] gave particularly interesting results, even though the accumulation of significant data was difficult, due to the already discussed aspect concerning the low numbers of MN detectable in this cell type following a treatment, with respect to exposure of a somatic compartment. Interestingly however, the expected patterns were confirmed: in control preparations, where centromere-positive MN were about 1/3 of all MN observed, the large majority of them showed 3 or more telomeric sequences, and 1 MN with 6 telomeres carried 2 centromeric signals [34]. Centromere-negative MN as expected showed

1-2 visible telomeric signals. In preparations obtained from animals exposed to suspected clastogenic agents, a larger proportion of MN carrying 1 or 2 telomeres only, and lacking the centromeric sequence, was found with respect to the controls [34]; on the other hand, preparations obtained after treating animals with the known aneuploidogenic agent chloral hydrate lead to a relative increase of centromere-positive MN with 2-8 telomeric sequences (Figure 7). It must be noticed in addition that the centromere-positive MN proportion detected with this approach in control preparations [34] is in very good agreement with that detected by tandem labelling (Figure 4) and with data coming from CREST [35] or FISH analysis [36] in the same cell type.

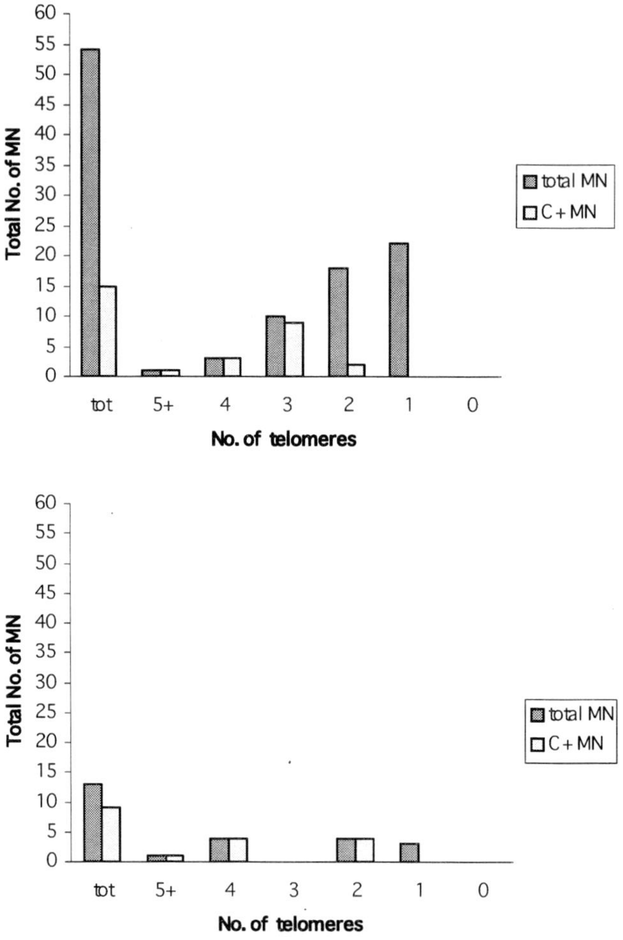

Figure 7. The application of the simultaneous detection of centromeric and telomeric sequences in early spermatids can contribute to make inferences on the mechanism of action of tested compounds in a cell system usually showing weak increases of MN with respect to the control. Top: distribution of telomeres in total and C+ MN in early spermatids of untreated mice, original data in [34]; bottom: distribution of telomeres in total and C+ MN in early spermatids of mice treated with chloral hydrate, a chemical with known aneuploidogenic action

Conclusion

Our data indicate that PRINS leads to accurate MN estimates, and that the sensitivity of this methodology is higher than FISH in dual colour approaches such as tandem labelling of centromeric and pericentromeric sequences [23, 29], or simultaneous detection of centromeric and telomeric sequences [16-17; 32-34]. The telomere labelling appeared particularly efficient with respect to the low sensitivity reported by studies applying the FISH methodology [16-17]. The highest performance of PRINS with respect to FISH in the MN assay is certainly due to the very low background noise, which allows the discrimination of true fluorescent signals from false positive also in the case of the small telomeric sequences.

The CREST approach consists in one day protocol (as well as for PRINS) and its advantage in the MN assay is represented by the fact that kinetochore proteins are widely conserved and kinetochores of several species can be recognised by the same antibodies [15]. In a direct comparison of the two methodologies, we have found that both of them are sensitive and accurate; PRINS however offers a clear advantage with respect to CREST analysis: because kinetochores must be well preserved to be immunodetected, many standard fixatives including those containing acetic acid, are not compatible with the technique [15, 30].

In conclusion, PRINS is a valid and fast approach which can be recommended to detect satellite DNA in application to the MN assay.

References

[1] Wright, E.G. (1999) Inherited and inducible chromosomal instability: a fragile bridge between genome integrity mechanisms and tumourigenesis. *Journal of pathology, 187,* 19-27.

[2] Little, J.B. (2000) Radiation carcinogenesis. *Carcinogenesis, 21,* 397-404.

[3] Ponsa, I., Barquinero, J.F., Mirò, R., Egozcue, J., Genescà, A. (2001) Non-disjunction and chromosome loss in gamma-irradiated human lymphocytes: a fluorescence *in situ* hybridation analysis using centromere-specific probes. *Radiation Research, 155,* 424-431.

[4] Heddle, J.A. (1973) A rapid in vivo test for chromosome damage. *Mutation Res., 18,* 187-193.

[5] Schmid, W. (1975) The micronucleus test. *Mutation Res., 31,* 9-15.

[6] Fenech, M. (2000) The in vitro micronucleus technique. *Mutation Res., 445,* 81-95.

[7] Kirsch-Volders, M., Sofuni, T., Aardema, M., Albertini, S., Eastmond, D., Fenech, M., Ishidate, M. Jr., Kirchner, S., Lorge, E., Morita, T., Norppa, H., Surralles, J., Vanhauwaert, A., Wakata, A. (2003) Report from the in vitro micronucleus assay working group. *Mutation Res., 540,* 153-163.

[8] Vig, B.K., Swearngin, S.E. (1986) Sequence of centromere separation: kinetochore formation in induced laggards and micronuclei. *Mutagenesis, 1,* 461-465.

[9] Degrassi, F. and Tanzarella, C. (1988) Immunofluorescent staining of kinetochores in micronuclei: a new assay for the detection of aneuploidy. *Mutation Res., 203,* 339-345.

[10] Hennig, U.G.G., Rudd, N.L., Hoar, D.I. (1988) Kinetochore Immunofluorescence in micronuclei: a rapid method for the *in situ* detection of aneuploidy and chromosome breakage in human fibroblasts. *Mutation Res., 203,* 405-414.

[11] Thomson, E.J. and Perry P.E. (1988) The identification of micronucleated chromosomes: a possible assay for aneuploidy. *Mutagenesis, 3,* 415-418.

[12] Eastmond, D.A. and Pinkel, D. (1990) Detection of aneuploidy and aneuploidy-inducing agents in human lymphocytes using fluorescence in situ hybridization with chromosome-specific DNA probes. *Mutation Res., 23,* 303-318.

[13] Migliore, L., Bocciardi, R., Macri, C., Lo Jacono, F. (1993) Cytogenetic damage induced in human lymphocytes by four vanadium compounds and micronucleus analysis by fluorescence in situ hybridization with a centromeric probe. *Mutation Res., 319,* 205-213.

[14] Norppa, H., Renzi, L., Lindholm, C (1993) Detection of whole chromosomes in micronuclei of cytokinesis-blocked human lymphocytes by antikinetochore staining and *in situ* hybridization. *Mutagenesis, 8,* 519-525.

[15] Schuler, M., Rupa, D.S., Eastmond, D.A. (1997) A critical evaluation of centromeric labeling to distinguish micronuclei induced by chromosomal loss and breakage in vitro. *Mutation Res., 392,* 81-95.

[16] Schriever-Schwemmer, G. and Adler, I.-D. (1994) Differentiation in micronuclei in mouse bone marrow cells: a comparison between CREST staining and fluorescence *in situ* hybridization with centromeric and telomeric DNA probes. *Mutagenesis, 9,* 333-340.

[17] Miller, B.M. and Nusse, M. (1993) Analysis of micronuclei induced by 2-chlorobenzylidene malonitrile (CS) using fluorescence in situ hybridization with telomeric and centromeric DNA probes, and flow cytometry. *Mutagenesis, 8,* 35-41.

[18] Masumoto, H., Masukata, H., Muro, Y., Nozaki, N., Okazaki. T. (1989) A human centromere antigen (CENP-B) interacts with a short specific sequence in alphoid DNA, a human centromeric satellite. *J. Cell Biol. 109,* 1963–1973.

[19] Lee, C., Wevrick, R., Fisher, R.B.,, Ferguson-Smith, M.A., Lin, C. C. (1997) Human centromeric DNAs, *Hum. Genet. 100,* 291–304.

[20] Mitchell, A.R. (1996) The mammalian centromere: Its molecular architecture. *Mutation Res., 372,* 153-167.

[21] Chen, H.W., Tomar,R., Eastmond.D.A. (1994) Detection of hydroquinone-induced non-random breakage in the centromeric heterochromatin of mouse bone marrow cells using multicolor fluorescence *in situ* hybridization with the mouse major and minor satelite probes. *Mutagenesis, 9,* 563-569.

[22] Rupa, D.S., Hasegawa, L., Eastmond, D.A. (1995) Detection of chromosomal breakage in the 1cen-q12 region of interphase human lymphocytes using multicolor fluorescence in situ hybridization with tandem DNA probes. *Cancer Res. 55,* 640-645.

[23] Renzi, L., F. Pacchierotti and A. Russo (1996) The centromere as a target for the induction of chromosome damage in resting and proliferating mammalian cells: assessment of mitomycin C-induced genetic damage at kinetochores and centromeres by a micronucleus test in mouse splenocytes. *Mutagenesis, 11,* 133-138.

[24] Eastmond, D.A., Rupa, D.S., Chen, H.W., Hasegawa L.S. (1993) Multicolor fluorescence in situ hybridization with centromeric DNA probes as a new approach to distinguish chromosome breakage from aneuploidy in interphase cells and micronuclei. In: B.K. Vig (Ed.), Chromosome Segregation and Aneuploidy. NATO 4SI Series, pp. 377-390.

[25] Eastmond, D.A., Schuler M., Rupa D.S. (1995) Advantages and limitations of using fluorescence in situ hybridization for the detection of aneuploidy in interphase human cells. *Mutation Res., 348,* 153-162.

[26] Krishna, G. and Hayashi, M. (2000) In vivo rodent micronucleus assay: protocol, conduct and data interpretation. *Mutation Res., 455,* 155-166.

[27] Russo, A. (2000) In vivo cytogenetics: mammalian germ cells. *Mutation Res., 445,* 167-189.

[28] Russo, A. (2006) Evaluation of chromosome instability in mammalian cells by PRINS: detection of repetitive DNA sequences in micronuclei. In: F. Pellestor (Ed.) PRINS and In Situ PCR Protocols, 2nd edition. Series "Methods in Molecular Biology", Humana Press TOTOWA, NJ, USA, pp 89-104.

[29] Russo, A., Tommasi, A.M., Renzi, L. (1996) Detection of minor and major satellite DNA in cytokinesis-blocked mouse splenocytes by a PRINS tandem labelling approach. *Mutagenesis, 11,* 547-552.

[30] Russo, A., Stocco, A., Majone, F. (1992) Identification of kinetochore containing (CREST"+") micronuclei in mouse bone marrow erythrocytes. *Mutagenesis, 7,* 195-197.

[31] Canova, S., Fiorasi, F., Mognato, M., Grifalconi, M., Reddi, E., Russo, A., Celotti, L. (2005) "Modeled microgravity" affects cell response to ionizing radiation and increases genomic damage. *Radiat Res., 163,* 191-199.

[32] Russo, A., Priante, G., Tommasi, A.M. (1996) PRINS localization of centromeres and telomeres in micronuclei indicates that in mouse splenocytes chromatid non-disjunction is a major mechanism of aneuploidy. *Mutation Res., 372,* 173-180.

[33] Basso, K. and Russo, A. (2000) Detection and characterization of micronuclei in a murine liver epithelial cell line, by application of the in vitro cytokinesis block MN assay and PRINS. *Mutagenesis, 15,* 349-356.

[34] Tommasi, A.M., de Conti, S., Dobrzynska M.M., Russo, A (1998) Evaluation and characterization of micronuclei in early spermatids of mice exposed to 1,3-butadiene. *Mutation Res. 397,* 45–54.

[35] Allen, J.W., Collins, B.W., Evansky, P.A. (1994) Spermatid micronucleus analyses of thrichloroetylene and chloral hydrate effects in mice, *Mutation Res. 323,* 81–88.

[36] Kallio, M. and Lahdetie J. (1993) Analysis of micronuclei induced in mouse early spermatids by mitomycin C, vinblastine sulphate or etoposide using fluorescence in situ hybridization. *Mutagenesis, 8,* 561–567.

In: PRINS and PNA Technologies in Chromosomal Investigations ISBN 1-60021-413-4
Editor: Franck Pellestor, pp. 69-77

Chapter 6

Chromosomal Analysis of Human Gametes by PRINS

***Franck Pellestor*[*1,2,3], *Brigitte Andréo*[2] *and Samir Hamamah*[1,3]**
[1] Department of Reproductive Biology B,
Arnaud de Villeneuve Hospital, Montpellier, France
[2] Institute of Human Genetics, CNRS UPR 1142, Montpellier, France
[3]Inserm, U 847, Montpellier, France

Abstract

The PRINS technique has made dramatic progress during the last decade and has moved from research benches to diagnostic laboratories. One of the most successful adaptation of the PRINS technique concerns the chromosomal investigation of human matures gametes. Several recent studies have demonstrated that PRINS was more efficient than FISH on human gametes. With the development of simplified protocol, PRINS has become an efficient tool for investigating the chromosomes in human gametes. The significant potential of the PRINS technique is attributable to the small size of the primers and their high specificity.

Key-words: PRINS, aneuploidy, spermatozoa, oocyte

Introduction

Chromosomal abnormalities are known to be the major reason for pre- and post-implantation human embryo wastage. Consequently, the direct chromosomal analysis of

---

[*] Correspondence to : Dr. Franck Pellestor, Department of Reproductive Biology B, Hôpital Arnaud de Villeneuve 371, avenue du Doyen Gaston Giraud, 34295 Montpellier Cedex 5, France. Tel: 33 (0)4 67 33 64 04; E-mail: Franck.Pellestor@igh.cnrs.fr

human gametes is an important tool for the investigation of the occurrence and the etiology of these abnormalities under a variety of clinical conditions.

Because human spermatozoa are easier to obtain than are mature human oocytes, most of the pertinent studies have focused on male gametes. Direct information on the chromosomal constitution of human sperm was first obtained with the introduction of the human sperm-hamster egg fertilization system, which allows the karyotyping of human spermatozoa [1, 2]. However, the technique is time consuming and labor intensive. It has taken 20 years to collect 20,000 sperm karyotypes from normal donors and 6,000 from men with constitutional chromosome aberrations [3, 4]

Data on the chromosomal constitution of human oocytes are much less abundant and harmonious. The development of in vitro fertilization (I.V.F.) programs has made human oocytes available, offering the opportunity to start cytogenetic investigations on in vitro recovered oocytes. Since 1984, more than 50 conventional cytogenetic surveys of human oocytes have been reported [5] but the results have shown wild variability in the incidence of abnormalities (ranging from 8% to 59%) and contradictions concerning the type of abnormalities [6] and etiological factors such as maternal age [7].

The introduction of molecular cytogenetic techniques for in situ cytogenetic analysis has largely contributed to stimulate studies on the chromosomal constitution of human gametes and aneuploidy occurrence. In addition, the direct in situ chromosomal analysis of isolated cells has constituted an important challenge, especially in conjunction with the clinical development of preimplantation genetic diagnosis (PGD).

Because of its relative simplicity and the commercial availability of numerous DNA probes and in situ hybridization kits, FISH has quickly become the standard technique for aneuploidy assessment in human gametes. To date, numerous FISH studies have been reported in both human sperm and human oocytes, providing a large body of data, but also pointing out the limitations of this technique, in terms of efficiency and specificity, when applied to human gametes [5, 8, 9]. Based on the in situ annealing of specific oligonucleotide primers and their extension by a Taq polymerase, the PRINS technique exhibits a high specificity for the identification and the discrimination of repeat DNA sequences [10, 11]. The high efficiency of PRINS labelling has been demonstrated in various types of cells [12, 13]. Since PRINS is efficient for chromosome detection and diagnostic purposes, the procedure has also been adapted to human gametes. Thanks to the development of simplified protocols, PRINS has become a powerful tool for investigating the chromosomes in both oocytes and spermatozoa.

Chromosome Detection by PRINS on Human Oocytes

The chromosomal constitution of human oocytes was first studied by conventional karyotyping, but the difficulties of obtaining good quality metaphase spreads and of performing accurate chromosome identification have strongly limited the fiability of this approach [5, 14]. The approximate nature of these studies was pointed out by Angell's observation that premature separation of homologous chromatids through anaphase I might be a significant mechanism for human aneuploidy [6]. The emergence of this particular mechanism of malsegregation introduced an important new parameter in the investigation of meiotic non-disjunction.

On human oocytes, molecular cytogenetic techniques, such as FISH and PRINS have appeared as a significant improvement over karyotyping because they overcame the difficulty of chromosome spreading and sometimes allowed the parallel analysis of the first polar body [15, 16]. However, it must be kept in mind that the information obtained was limited to the targeted chromosomes.

Only a few PRINS experiments have been performed on human oocytes to date. In 1996, the multicolour PRINS procedure was successfully tested on 11 human oocytes obtained from patients participating in an IVF program [17]. Oocytes were individually fixed according to the air-drying method of Tarkowski and PRINS labelling was performed, using specific centromeric primers for chromosomes 1, 8, 9, 12, 13, 16, 18 21, and X. Single-target PRINS reaction was extremely rapid. Some primers, such as chromosome 1 or chromosome 9 primers gave in situ signals in less than 15 minutes. On average, the labelling of one chromosome was achieved within 30 minutes. The fluorescent spots were well defined and easily scorable. The feasibility of multi-target PRINS labelling was tested with various combinations of primers. On this small series of 11 oocytes, two numerical abnormalities were found, involving a nullosomy for chromosome 16 and a disomy for chromosome 8. Another assay was performed in 1999 on 118 oocytes, using specific primers for chromosomes X and Y, in order to evaluate the reliability of PRINS procedure for PGD [18].

Recently the new ultra-fast multicolour PRINS protocol introduced by by Yan et al. [19] has been tested on human oocytes and polar bodies. On the basis of the direct in situ mixing of several fluorochromes incorporated in sequential PRINS reactions, this PRINS protocol allows rapid and efficient labelling of 3 or 4 individual chromosomes. Thus, various combinations of 3 or 4 chromosome-specific primers (for chromosome 1, 7, 9, 16, and 18) were used on a sample of 16 oocytes and 7 polar bodies [20]. These reactions were carried out with 3 distinct fluorochrome-dUTPS: fluoresceine isothiocyanate (FITC-11-dUTP), tetramethylrhodamine isothiocynate (TRITC-6-dUTP), and Cascade Blue-7-dUTP. On this material, each PRINS reaction required 6 minutes for optimal labelling of the chromosome targets. The 3 colour experiments resulted in well distinct coloured spots in all the 10 oocytes and in 3 corresponding polar bodies. The presence of a supernumerary single chromatid was observed in one oocyte and in one case, the balanced separation of the 2 chromatids of chromosome 18 was observed. The 4 colour experiments, using the combination FITC/TRITC/FITC/CascadeBlue, gave satisfactory results in all cases, i.e. the presence of clearly distinct and unique 4 well-coloured signals on metaphases and polar bodies (Fig. 1). One case of supernumerary chromatid for chromosome 16 was observed, as well as the balanced separation of the 2 chromatids of chromosome 16.

This new protocol significantly simplifies the multicolour PRINS procedure and its adaptation to isolated cells. Using the direct incorporation of fluorochromes, these experiments indicated that a ratio of colour must be respected to get well-defined green, red and yellow signals. The 3 distinct colours are only obtained when the final proportion of green fluorophore is twice the rate of the red one.

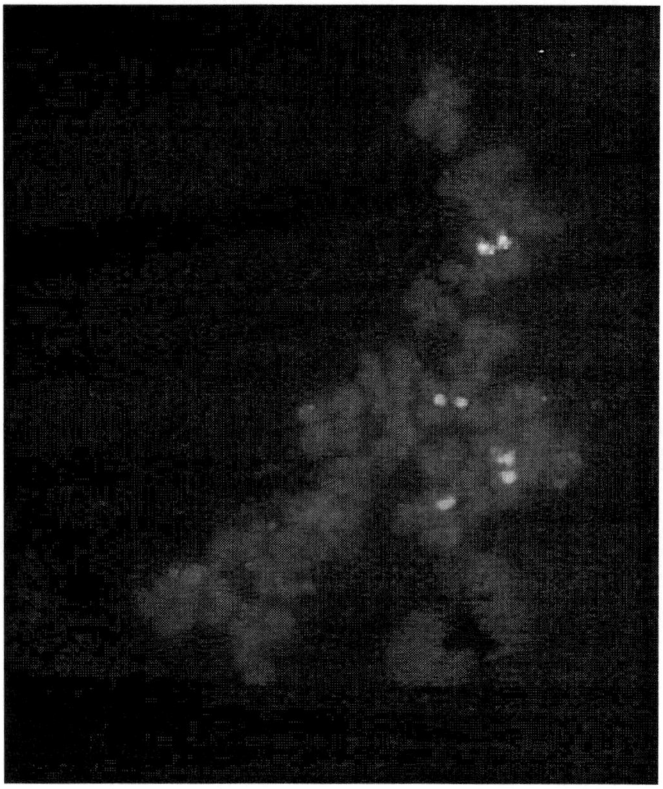

Figure 1. Four color labeling of human oocyte chromosomes using the labeling order FITC/TRITC/FITC/Cascade Blue. Chromosome 1 is labeled in yellow, chromosome 7 in green, chromosome 9 in red and chromosome 16 in blue.

Such a difference in the fluorescence intensity seems to be linked to the intrinsic characteristics of the 3 fluorophores used, fluorescein, rhodamine and Cascade Blue. Paradoxically, although FITC is one of the most popular fluorescent reagents, this dye presents some features, which significantly limit its in situ utility for multicolour application, such as a high rate of photobleaching and a significant pH sensitivity. TRITC is more photostable, not pH sensitive and less quenchable, and this constitutes one of the most efficient fluorophores, which always appears brighter than the corresponding fluorescein conjugates [21]. The Cascade Blue dye displays only a little spectral overlap with fluorescein and provides a colour, which contrasts well with the longer-wavelength green and red fluorophores [22]. In addition, because Cascade Blue is used for the last PRINS reaction in the sequential procedure, the graduated saturation of the alpha-satellite sequences of the other chromosomes labelled by the previous PRINS reactions, must limit the incorporation of the blue labelled nucleotide. Thus, the resulting signals remain clearly distinguishable.

These preliminary results demonstrated the efficiency of the new multicolour PRINS procedure using the most currently used fluorophores. The availability of new distinct dyes with narrow spectral bandwidths and good photostability, such as the cyanines and the AlexaFluors available in a large variety of fluorescent colours, could permit to use different and perhaps larger combinations of fluorochromes for the in situ detection of more than 4 chromosomes by PRINS.

Chromosome Detection by PRINS on Human Sperm

In situ labelling with chromosome-specific DNA probes has offered a new approach for investigating the chromosomal constitution of human spermatozoa. Aneuploidy has been estimated for several chromosomes using centromeric repeat probes in FISH assays, but a review of the literature indicates that among FISH studies, there are significant variations in the reported estimates of disomy for individual chromosomes [8, 9]. The use of centromeric probes presents certain limitations because several human chromosomes have similar sequences in their satellite DNA. This results in cross-hybridization in FISH reaction. FISH analysis of spermatozoa is also hampered by the fact that the DNA in sperm heads is tightly condensed and difficult to access. A decondensing step is needed to make the DNA sequences accessible to the hybridizing probes, and the success of FISH is largely dependent on the efficiency of the decondensing treatment. However, there is no consensus concerning the application of this procedure on sperm nuclei.

The adaptation of PRINS technique to human male gametes constituted a very interesting challenge because of the particularities of sperm nucleus in term of genomic compaction and accessibility of DNA sequence. Thus, the PRINS methodology has been combined with the NaOH treatment of sperm preparations, which allows the simultaneous decondensation and denaturation of sperm nuclei [23]. In PRINS reaction, the decondensation of the sperm head is a less limiting factor than in FISH because of the small size of the oligonucleotide primers. This greatly facilitates their penetration into sperm nuclei and their access to the genomic target sequences, resulting in a more efficient and rapid labelling of sperm nuclei. This might explain the extreme rapidity of PRINS labelling of human sperm. Indeed, a single target PRINS reaction can be performed in 10 minutes. The time of optimal NaOH treatment depends on the age of the sperm preparation slides and the concentration of NaOH solution. Initially, we used 3M NaOH, but numerous experiments and a comparison of the results in terms of quality of the preparation obtained led us to adopt a 0.5M NaOH solution. Best results in labelling efficiency were obtained with 2-day-old slides and 5 minutes 0.5 M NaOH pretreatment. The NaOH treatment induces uniform swelling of the sperm nucleus till 1.5 to 2 times its normal size, and maintains the characteristics and shape of the sperm nucleus, including the tail. This allows the differentiation between spermatozoa and other cells such as leukocytes or immature germ cells present in the ejaculate.

In practice, fresh sperm samples were first sample was washed twice in 1xPBS by centrifugation (5 min at 2000 rpm) and fixed for 1 hour in fresh fixative (3:1 methanol:glacial acetic acid) at 4°C. The sperm suspension was then dropped onto clean microscope slides and air-dried. Slides were aged for 2 day at room temperature, and just before the PRINS procedure, the slides were immersed in a 0.5 M NaOH solution at room temperature for 5 min, passed through an ethanol series (70%, 90%, 100%) and air-dried.

Double and triple PRINS reactions have been performed using various combinations of chromosome-specific primers. Both the efficiency and the reliability of the PRINS technique on sperm were tested, by using different primers to label individual chromosomes. Using this PRINS protocol, we have directly estimated diploidy and disomy frequencies for 15 autosomes and sex chromosomes in sperm samples from several normal fertile donors [24].

The labelling efficiency usually ranges from 97 to 100% according to the primer tested. A minimum of 10,000 nuclei per chromosome was analyzed. For the autosomes, the mean disomy rate ranged from 0.18% to 0.36 %. Among all donors, chromosome 21 exhibited the highest disomy rate (from 0.30% to 0.39%) suggesting that in male meiosis, chromosome 21 may be more susceptible to nondisjunction than the other autosomes. The frequency of disomy for gonosomes ranged from 0.08% to 0.13%. Inter-individual variations in disomy rates were observed but these were statistically non significant. Inter-individual heterogeneity was more evident in diploidy where frequencies ranged from 0.08% to 0.45% among the normal subjects tested.

The efficiency of the new multicolour PRINS protocol, based on the direct mixing of the colour of fluorochromes, was also tested and validated on human sperm [25]. Comparative estimates of disomy were performed for several chromosomes on sperm samples from 2 donors, either by the new 3-colour protocol (Fig. 2) or by the conventional dual colour PRINS procedure previously described. There was no statistical difference between the disomy rates obtained with the conventional dual colour PRINS technique and the fast 3-colour procedure. This new protocol provides significant simplifications in the multicolour PRINS protocol without modifying the efficiency and the specificity of the labelling reaction. The use of this multicolour protocol has advantages of rapidity and low cost as compared to FISH.

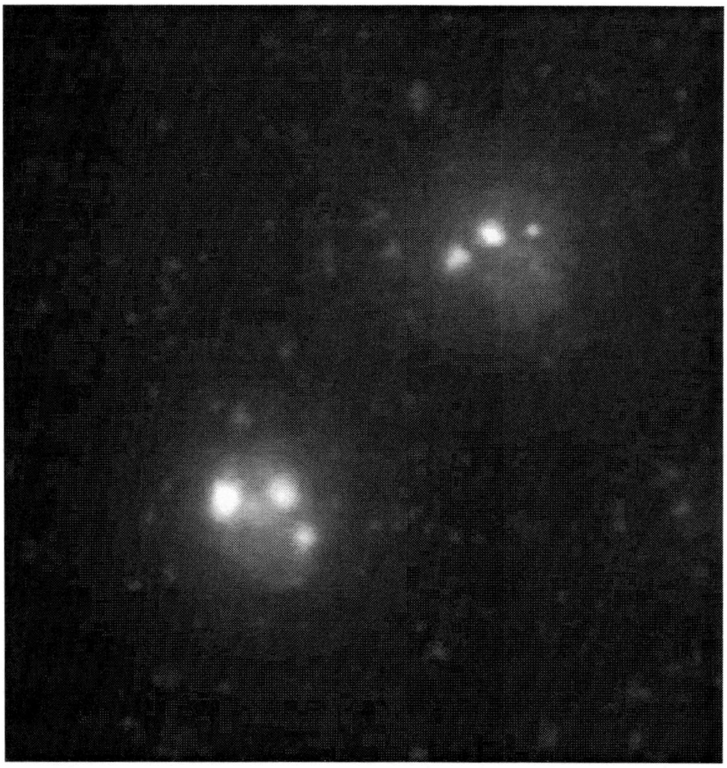

Figure 2. Direct PRINS labelling of human sperm nuclei using primers for chromosomes 1 (yellow), 9 (red) and 7 (green).

The PRINS procedure has also been used to directly investigate in sperm the meiotic segregation patterns of reciprocal translocations [26] and to investigate the occurrence of inter-chromosomal effect in sperm of chromosomal rearrangement carriers [27].

Conclusion

The PRINS reaction combines the sensitivity of the polymerase chain reaction with the cytological localization of specific DNA sequences, and display several features that make it efficient for chromosome detection, such as the rapidity of the reaction, the high specificity of the labelling but also the low cost of the reaction. Since its introduction, the PRINS technique has quickly evolved from basic research to diagnosis procedures, and its successful use on human gametes has proven that this technique could be used on difficult biological material, and consequently it has a great potential for clinical applications.

Acknowledgments

This study was supported by a European INTAS project (N° 03-51-4060).

References

[1] Rudak, E., Jacobs, P.A., and Yanagimachi, R. (1978) Direct analysis of the chromosome constitution of human spermatozoa. *Nature* 271, 911-913.

[2] Martin, R.H., and Rademaker, A.W. (1987) The effect of age on the frequency of sperm chromosomal abnormalities in normal men. *Am. J. Hum. Genet.* 41, 484-492.

[3] Pellestor, F. (1991) Differential distribution of aneuploidies in human gametes according to their sex. *Hum. Reprod.* 6, 1252-1258.

[4] Guttenbach, M., Engel, W., and Schmid, M. (1997) Analysis of structural and numerical chromosome abnormalities in sperm of normal men and carriers of constitutional chromosome aberrations. A review. *Hum. Genet.* 100, 1-21.

[5] Pellestor, F., Anahory, T., and Hamamah, S. (2005) The chromosomal analysis of human oocytes. An overview of established procedures. *Hum. Reprod. Up.* 11, 15-32.

[6] Angel, R.R. (1997) First-meiotic-division nondisjunction in human oocytes. *Am. J. Hum*. Genet. 61, 23-32.

[7] Pellestor, F., Andréo, B., Arnal, F., Humeau, C., and Demaille, J. (2003) Maternal aging and chromosomal abnprmalities: new data drawn from in vitro unfertilized human oocytes. *Hum. Genet.* 112, 195-203.

[8] Downie, S.E., Flaherty, S.P., and Matthews, C.D. (1997) Detection of chromosomes and estimation of aneuploidy in human spermatozoa using fluorescence in-situ hybridization. *Mol. Hum. Reprod.* 3, 585-598.

[9] Egozcue, J., Blanco, J., and Vidal, F. (1997) Chromosome studies in human sperm nuclei using fluorescence in-situ hybridization (FISH). *Hum Reprod. Update.* 3, 441-452.

[10] Pellestor, F., Girardet, A., Andréo, B., and Charlieu, J.P. (1994) A polymorphic alpha satellite sequence specific for human chromosome 13 detected by oligonucleotide primed in situ labeling (PRINS). *Hum. Genet.* 94, 346-348.

[11] Serakinci, N., and Koch, J. (1999) detection and sizing of telomeric repeat DNA in situ. *Nat. Biotechnol.*17, 200-201.

[12] Hindkjaer, J., Koch, J., Terkelsen, C., Brandt, C.A., Kolvraa, S., and Bolund, L. (1994) Fast, sensitive, multicolour detection of nucleic acids in situ by primed in situ labeling (PRINS). *Cytogenet. Cell Genet.* 66, 152-154.

[13] Orsetti, B., Lefort, G., Boulot, P., Andréo, B., and Pellestor, F. (1998) Fetal cells in maternal blood: the use of primed in situ (PRINS) labelling technique for fetal cell detection and sex assessment. *Prenat. Diagn.* 18, 1014-1022.

[14] Pellestor, F., Andréo, B., Arnal, F., Humeau, C., and Demaille, J. (2002) Mechanisms of non-disjunction in human female meiosis: the co-existence of two modes of malsegregation evidenced by the karyotyping of 1397 in-vitro unfertilized oocytes. *Hum. Reprod.* 17, 2134-2145.

[15] Dailey, T., Dale, B., Cohen, J., and Munné, S. (1996) Association between nondisjunction and maternal age in meiosis-II human oocytes. *Am. J. Hum. Genet.* 59, 176-184.

[16] Kuliev, A., Cieslak, J., Ilkevitch, Y., and Verlinsky, Y. (2003) Chromosomal abnormalities in a series of 6,733 human oocytes in preimplantation diagnosis for age-related aneuploidies. *Reprod. Biomed. Online* 6, 54-59.

[17] Pellestor, F., Girardet, A., Lefort, G., Andréo, B., and Charlieu, J.P. (1996) Rapid chromosome detection in human gametes, zygotes, and preimplantation embryos using the PRINS technique. J. *Ass. Reprod. Genet.* 13, 675-680.

[18] Findlay, I., Corby, N., Rutherforf, A., and Quirke, P. (1998) Comparison of FISH, PRINS, and conventional and fluorescent PCR for single-cell sexing: suitability for preimplantation genetic diagnosis. *J. Assist. Reprod. Genet.*15, 258-265.

[19] Yan, J., Bronsard, M., and Drouin, R. (2001) Creating a new color by omission of 3 end blocking step for simultaneous detection of different chromosomes in multi-PRINS technique. *Chromosoma* 109, 565-570.

[20] Pellestor, F., Anahory, T., Andréo, B., Régnier-Vigouroux, G., Soulié, J.P., Baudouin, M., and Demaille, J. (2004) Fast multicolour primed in situ protocol for chromosome identification in isolated cells may be used for human oocytes and polar bodies. *Fertil. Steril.* 81, 408-415.

[21] Kasten F.H. (1993) Introduction to fluorescent probes: properties, history and applications. In: Masson WT, ed. *Fluorescent and luminescent probes for biological activity.* New York: Academic Press, pp 12-33.

[22] Haugland, R.P. (1996) *Handbook of fluorescent probes and research chemicals.* Leiden, The Netherlands: Molecular Probes.

[23] Pellestor, F., Girardet, A., Coignet, L., Andréo, B., and Charlieu, J.P. (1996) Assessment of aneuploidy for chromosomes 8, 9, 13, 16 and 21 in human sperm by using primed in situ labeling technique. *Am. J. Hum. Genet.* 58, 797-802.

[24] Pellestor, F., Imbert, I., and Andréo, B. (2002) Rapipd chromosome detection by PRINS in human sperm. *Am. J. Med. Genet.* 107, 109-114.

[25] Pellestor, F., Malki, S., Andréo, B., and Lefort, G. (2002) Ultra-rapid multicolor PRINS protocol for chromosome detection in human sperm. *Chromosome Res.* 10, 359-367.

[26] Pellestor, F., Girardet, A., Coignet, L., Andréo, B., Lefort, G., and Charlieu, J.P. (1997) Cytogenetic analysis of meiotic segregation in sperm from two males heterozygous for reciprocal translocations using PRINS and humster techniques. *Cytogenet. Cell Genet.* 78, 202-208.

[27] Pellestor, F., Imbert, I., Andréo, B., and Lefort, G. (2001) Study of the occurrence of interchromosomal effect in spermatozoa of chromosomal rearrangement carriers by fluorescence in-situ hybridization and primed in-situ labelling techniques. *Hum. Reprod.* 16, 1155-1164.

In: PRINS and PNA Technologies in Chromosomal Investigations ISBN 1-60021-413-4
Editor: Franck Pellestor, pp. 79-89

Chapter 7

PRINS Applications in Plant Cytogenetics and Genomics

***Jiří Macas*[*1], *Alice Navrátilová*[1], *Marie Kubaláková*[2] and *Jaroslav Doležel*[2]**

[1] Biology Centre ASCR, Institute of Plant Molecular Biology, Branišovská 31, České Budějovice, Czech Republic

[2] Institute of Experimental Botany, Sokolovská 6, Olomouc, Czech Republic

Abstract

Shortly after the application of PRINS in human and animal cytogenetics, the method was adapted for plant materials, offering a simple and rapid alternative to fluorescence in situ hybridization. Here we provide a brief overview of PRINS applications in plant cytogenetics, with emphasis on investigation of chromosome distribution, long-range organization, and evolution of repetitive DNA sequences. Examples also include using PRINS for physical genome mapping and in chromosome flow cytometry and sorting, where it has been routinely employed to identify flow-sorted chromosomes in a number of species. On the other hand, PRINS was found of only limited use in flow cytometric discrimination of morphologically similar chromosomes, most likely due to non-quantitative nature of the labeling.

Keywords: Plant chromosomes; repetitive sequences; satellite DNA; PRINS; flow cytometric sorting

---

[*] Corresponding author: Jiří Macas, Institute of Plant Molecular Biology, Branišovská 31, České Budějovice, CZ-37005, Czech Republic, e-mail: macas@umbr.cas.cz; Phone: +420 387775516; Fax: +420 385310356

Introduction

Despite the enormous progress in genome sequencing techniques, molecular-cytogenetic approaches remain indispensable in investigating long-range molecular organization of plant genomes and their functional parts, the chromosomes. The techniques for in situ localization of specific DNA sequences on nuclei, chromosomes, and chromatin fibers provide valuable information during all stages of plant genome mapping, from discriminating individual chromosomes within the karyotypes (Cuadrado and Jouve 1997; Navrátilová et al. 2003), to assembling or verifying contigs of clones spanning centromeric or other repeat-rich regions in sequenced genomes (Cheng et al. 2001; Nagaki et al. 2004). These methods are often based on fluorescence in situ hybridization (FISH), a technique employing hybridization and subsequent fluorescent detection of labeled DNA or RNA probes (Leitch et al. 1994).

In contrast to FISH, primed in situ labeling (PRINS) employs unlabeled probes which upon annealing to their target sequences within chromosomal or nuclear DNA serve as primers for incorporation of labeled dNTPs by a DNA polymerase. Thus, the reaction conditions are very similar to those used in polymerase chain reaction (PCR), including cycles of template denaturation, primer annealing and extension, which together result in sequence-specific deposition of labeled nucleotides. Similarly to PCR, various thermostable DNA polymerases are the enzymes of choice for PRINS. However, often only a single reaction cycle is sufficient to achieve the desired labeling of repetitive DNA targets. Typically, one or more synthetic oligonucleotides are used as primers, although successful labeling was also achieved using restriction enzyme-digested DNA fragments (Abbo et al. 1993b; Macas et al. 1995).

Following the development of PRINS and its application to sequence detection on human chromosomes (Koch et al. 1989; Gosden et al. 1991), the method was also successfully introduced to plant cytogenetics (Abbo et al. 1993a, 1993b; Macas et al. 1995; Thomas et al. 1996). PRINS in plants follows the methodology developed for human and animal material, except for modifications to chromosome preparations. Squash preparations of root tip meristems are preferred, but the reaction was also performed using isolated chromosomes either in suspension or after flow cytometric sorting onto microscope slides (Macas et al. 1995; Pich et al. 1995; Lysák et al. 1999; Kubaláková et al. 2000). Several variations of the basic PRINS protocol have been introduced in order to eliminate non-specific background signal arising from the extension of free 3' DNA ends occurring at random nicks or gaps within chromosomal DNA. The intensity of this nonspecific labeling depends on the degree of DNA damage during chromosome isolation, the age of chromosome preparations, and the plant species (Kubaláková et al. 2000), and it can be at least partially eliminated by repairing the nicks using DNA ligase prior to PRINS (Kubaláková et al. 1997). Alternatively, the free 3' ends can be blocked by enzymatic incorporation of dideoxynucleotides, which prevent their extension in the subsequent PRINS reaction (Menke et al. 1998; Macas et al. 2006b). As this treatment improves the signal-to-background ratio, it is especially useful for detection of less abundant targets that could be otherwise obscured. However, it is not usually necessary for detection of repetitive DNA targets.

In addition to elimination of the background signal, the sensitivity of PRINS can be enhanced by using various signal amplification systems employing fluorescently labeled antibodies instead of direct visualization of fluorochrome-conjugated nucleotides. For example, using digoxigenin-labeled nucleotides and their detection by anti-DIG-fluorescein antibody was shown to improve the reaction sensitivity (Macas et al. 1995; Menke et al. 1998). Although a single reaction cycle is sufficient for labeling abundant targets, PRINS sensitivity can be further increased by performing repeated cycles of DNA denaturation, primer annealing and extension (cycling- or C-PRINS, PCR-PRINS). Signal amplification by C-PRINS was reported for low-copy (Zhu et al. 1995; Shi et al. 1996) as well as for repetitive sequences (Kubaláková et al. 1997; Kubaláková and Doležel 1998), although amplification during consecutive cycles was not exponential as in the case of PCR.

There are several reports of detection of low-copy genomic sequences on plant chromosomes, including localization of the B-hordein gene cluster on barley chromosome 1H (Abbo et al. 1993a), and detection of RFLP loci in soybean (Zhu et al. 1995; Shi et al. 1996). However, PRINS sensitivity in most plant species is generally lower compared to FISH and does not permit routine visualization of single-copy targets (Menke et al. 1998). On the other hand, it is comparable, if not superior, to FISH in terms of speed and simplicity. For example, efficient labeling of FokI satellite repeats in *Vicia faba* can be achieved in a total reaction time of only 7 min (Kubaláková et al. 1997). Thus, PRINS is most often used for rapid detection of repetitive DNA sequences on plant chromosomes, for discrimination of sequence variants of repeat subfamilies, and as a tool for chromosome identification in flow-cytometry. These applications are described in the following sections and summarized in Table 1.

Table 1. Summary of published reports on the use of PRINS in plants

Species	Target	Primer	Notes	References
Avena sativa	telomere	oligo		Thomas et al. 1996
Glycine max	RFLP loci	oligo	PCR-PRINS	Zhu et al. 1995; Shi et al. 1996
Hordeum vulgare	B-hordein gene cluster	oligo		Abbo et al. 1993a
	telomere	oligo		Thomas et al. 1996
	HvT01 tandem repeat	oligo		Menke et al. 1998
	microsatellite (GAA)n	oligo	C-PRINS	Lysák et al. 1999; Kubaláková et al. 2000
Lolium x Festuca	telomere	oligo		Thomas et al. 1996
Pisum sativum	rDNA	cloned rDNA digested with *Hpa*I and *Taq*I, or oligo	PRINS in suspension	Macas et al. 1995
	5S rRNA genes	oligo	C-PRINS	Kubaláková et al. 1997
	telomere	oligo	C-PRINS	Kubaláková et al. 1997
Secale cereale	rDNA	cloned rDNA digested with *Sau*3AI		Abbo et al. 1993b

Table 1. Continued

Species	Target	Primer	Notes	References
	B-chromosome specific repeat D1100	nested oligos	C-PRINS	Kubaláková et al. 1997
	rDNA, 5S rRNA genes	oligo		Mukai and Appels 1996
Trifolium repens	telomere	oligo		Thomas et al. 1996
Triticale	telomere	oligo		Thomas et al. 1996
Triticum aestivum	microsatellite (GAA)n, Afa satellite	oligo	C-PRINS	Vrána et al. 2000; Kubaláková et al. 2000, 2002
Vicia faba	FokI satellite	oligo		Menke et al. 1998; Kubaláková et al. 1997, 2000
		oligo	PRINS-FISH	Kubaláková et al. 1998; Nouzová et al. 1999
		oligo	PRINS in suspension, flow-sorting of chromosomes based on PRINS signals	Macas et al. 1995; Pich et al. 1995
	rDNA	oligo		Menke et al. 1998
	5S rRNA genes	oligo		Menke et al. 1998
	telomere	oligo		Menke et al. 1998
		oligo	C-PRINS	Kubaláková et al. 1997, 1998
	pVf7 (IGS-like repeat)	oligo		Menke et al. 1998
Vicia grandiflora	VicTR-B satellite	oligo	PRINS-FISH	Navrátilová et al. 2003
		oligo	discrimination of sequence subfamilies	Macas et al. 2006a; Macas et al. 2006b
Vicia pannonica	VicTR-A satellite	oligo		Macas et al. 2000
		oligo	PRINS-FISH	Navrátilová et al. 2003
Vicia sativa	VicTR-B satellite	oligo		Macas et al. 2000
	IGS-like satellite	oligo	discrimination of sequence subfamilies	Macas et al. 2003

Rapid and Highly Specific Detection of Repetitive DNA Sequences

Successful PRINS labeling was reported for a variety of plant repeats with tandem genomic organization. They include sequences present across different plant genomes as genes encoding ribosomal RNAs (Abbo et al. 1993b; Macas et al. 1995; Mukai and Appels 1996; Kubaláková et al. 1997; Menke et al. 1998), telomeric repeats (Thomas et al. 1996; Kubaláková et al. 1997; Menke et al. 1998), or microsatellites (Lysák et al. 1999; Kubaláková et al. 2000; Vrána et al. 2000; Doležel et al. 2004), as well as genus- or species-specific satellite repeats (Macas et al. 2000; Macas et al. 2003). Some of these sequences produced labeling patterns that allowed discrimination of morphologically similar chromosome types and could thus be used for karyotype descriptions (Macas et al. 2000; Navrátilová et al. 2003). Chromosome identification using PRINS labeling of satellite repeats was also employed in a modified protocol combining PRINS with fluorescence in situ hybridization (PRINS-FISH, Kubaláková et al. 1998). This two-step protocol consists of PRINS labeling of abundant satellite repeats providing cytogenetic landmarks, followed by FISH detection of the probe of interest labeled by a different fluorochrome. As the chromosomal DNA remains denatured after the completion of the PRINS reaction, the preparations can be used directly for FISH without any pretreatment. This approach was employed for simultaneous FISH localization of the repeated sequence TIII15 and the PRINS-labeled FokI repeats on field bean chromosomes (Nouzová et al. 1999), and for chromosome karyotyping in four other *Vicia* species using a combination of several repetitive probes (Navrátilová et al. 2003).

In addition to rapid visualization of various DNA repeats, PRINS can also be used for discrimination and selective labeling of specific repeat variants (sequence subfamilies) by employing primers targeting their characteristic mutations. This application takes full advantage of the specific features of the PRINS reaction, because using unlabeled primers allows cost-efficient testing of a number of primers which are much cheaper compared to labeled oligonucleotide probes required for FISH. Moreover, the primer specificity is preferentially determined by the complementarity of its 3' end to the target sequence, in the same manner as in PCR. Since PRINS buffers and other reaction conditions are the same or very similar to those used for PCR, it is possible to test a primer's specificity in PCR-based assay using cloned template sequences prior to its application in PRINS (Macas et al. 2006a). In plants, specific PRINS detection of sequence variants was reported for rRNA genes of rye (Mukai and Appels 1996) and for centromeric satellites in Arabidopsis (Heslop-Harrison et al. 1999). Recently, this approach was used for discrimination of *Vicia grandiflora* VicTR-B repeat subfamilies differing by only a few nucleotides (Macas et al. 2006a; Macas et al. 2006b). It was also employed for selective labeling of IGS-like repeat subfamilies in *Vicia sativa*, and revealed their differential amplification on individual chromosomes (Macas et al. 2003). As it is demonstrated in Figure 1, a specific repeat subfamily can be readily discriminated from the consensus sequence using only a single primer targeted to a variable sequence domain.

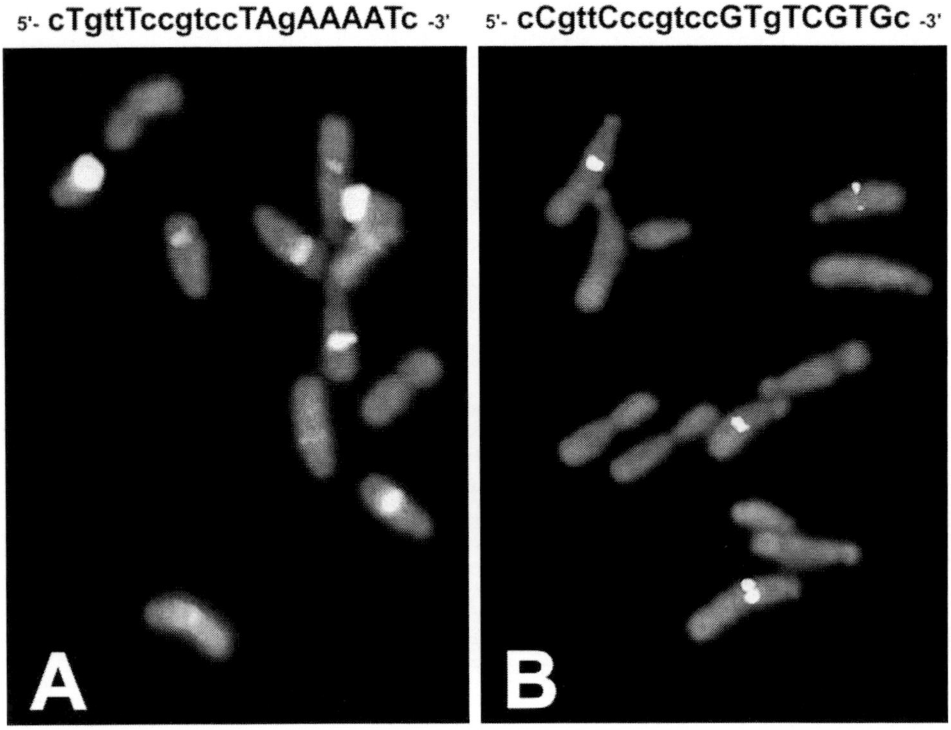

Figure 1. PRINS discrimination of satellite repeat subfamilies in *Vicia sativa*. The family of IGS-like satellites is derived from tandem subrepeats occurring within the 25S-18S rDNA intergenic spacer. It includes several sequence variants (subfamilies) which are differentially amplified on individual *V. sativa* chromosomes (Macas et al. 2003). All these loci can be detected using PRINS primer corresponding to the prevailing sequence variant (A). On the other hand, only a subset of the repeat loci is labeled using a primer targeted to the subsequence characteristic for the subfamily S12-4 (B). IGS-like repeats were labeled using direct PRINS incorporation of Alexa-fluor-488-dUTP (Molecular Probes) and appear yellow-green. The chromosomes were counterstained with DAPI and are displayed in red color. Primer sequences are shown on top, the nucleotides which differ between the primers are shown in capital letters

Chromosome Flow Cytometry and Sorting

During the last decade, chromosome analysis and sorting using flow cytometry (flow cytogenetics) has become an important tool of plant genetics and genomics as it allows isolation of particular chromosomes or chromosome arms in large quantities and high purities. Such dissection of complex genomes into smaller parts facilitates their efficient analysis and speeds up physical mapping and sequencing efforts (Doležel et al. 2005). Until now, flow cytogenetics has been developed for more than fifteen plant species, including economically important cereals and legumes (Doležel et al. 2004). Chromosome classification using flow cytometry results in a distribution of relative DNA content, or a flow karyotype, whose characterization requires assignment of chromosomes to individual peaks (Figure 2).

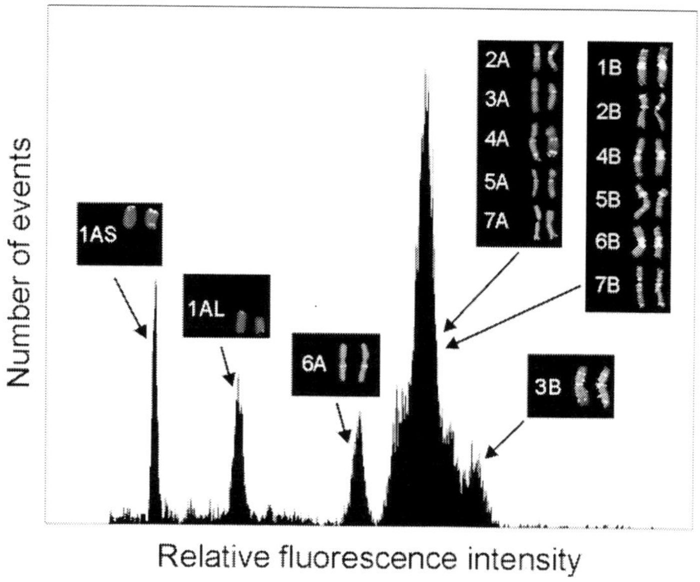

Figure 2. Identification of flow-sorted chromosomes using PRINS with microsatellite primers. Flow karyotype (histogram of relative fluorescence intensities) obtained after the analysis of DAPI-stained chromosome suspension prepared from double ditelosomic line of durum wheat cv. 'Langdon' (2n=26+2t1AL+2t1AS). The flow karyotype consists of a large composite peak and four small peaks. The composite peak represents a group of 11 chromosomes; small peaks represent chromosomes: 1AS, 1AL, 6A and 3B (from left to right). With the exception of 1AL and 1AS, chromosome content of individual peaks was determined after PRINS on chromosomes sorted onto microscope slide using a pair of primers for (GAA)n microsatellites and fluorescein-labeled nucleotides (yellow-green color). Chromosomes were counterstained with propidium iodide (red color). Due to a specific labeling pattern, each of the chromosomes can be unambiguously identified and the purity of sorted fractions can be estimated. Telocentric chromosomes 1AL and 1AS were identified by dual color FISH with probes for the *Afa* repeat (yellow color) and telomeric repeat (red color). The chromosomes were counterstained with DAPI (blue color). While the localization of the *Afa* repeat facilitated chromosomes identification, the use of telomeric probes was used to confirm that intact telocentric chromosomes were sorted

Moreover, a method is needed for unambiguous chromosome identification in order to assess the purity of sorted chromosome fractions. These tasks are best achieved by microscopic analysis of chromosome morphology and labeling patterns after visualization of selected repetitive sequences. Due to their speed and simplicity, PRINS and C-PRINS appeared to be attractive alternatives to FISH for routine labeling of repetitive DNA on sorted chromosomes (Kubaláková et al. 2000). Indeed, these methods were used to identify flow-sorted chromosomes and determine purity of flow-sorted fractions in a number of species including garden pea (Gualberti et al. 1996), barley (Lysák et al. 1999), field bean (Kubaláková et al. 2000), and wheat (Vrána et al. 2000, Kubaláková et al. 2002).

A chromosome can be sorted only if its relative DNA content or nucleotide composition differs significantly from all other chromosomes within the complement. This requirement complicates the development of plant flow cytogenetics, as in most species chromosomes are similar in size and AT/GC ratio and thus cannot be resolved (Lucretti and Doležel 1997; Kovářová et al. 2007). Consequently, the attention was turned to labeling repetitive DNA sequences in order to provide specific fluorescent tags allowing further chromosome discrimination. As the labeling needs to be performed on chromosomes in liquid suspension, a PRINS-based approach was preferred over FISH, because it allowed development of a simple procedure minimizing chromosome loss by reducing chromosome centrifugation steps. The procedure termed PRINSES (PRINS *en* suspension) was optimized using chromosome suspensions from garden pea and field bean and using ribosomal DNA sequences and FokI repeat elements as targeted repeats (Macas et al. 1995). In order to avoid precipitation of cellular proteins leading to chromosome clumping, crude chromosome suspensions were purified on sucrose gradients prior to PRINSES. Under optimal conditions, strong and specific signals were obtained exclusively on chromosome regions known to contain the respective DNA sequences. The assumption that fluorescent labeling of specific DNA repeats by PRINSES could aid in discrimination of individual chromosomes by flow cytometry was confirmed by Pich et al. (1995) using field bean as a model species. The authors analyzed chromosomes from a translocation line ACB with a reconstructed karyotype, where monoparametric analysis after DAPI staining resolved only the smallest and the largest chromosome (Doležel and Lucretti 1995). On the other hand, bivariate analysis after staining chromosomal DNA by propidium iodide and fluorescently labeling FokI repeats by PRINSES resolved all six chromosomes of the complement (Pich et al. 1995). It should be noted, however, that recent attempts to employ this technique for chromosome discrimination and sorting in barley, garden pea and wheat failed, in spite of successful labeling of various repeats on chromosomes in suspension (Doležel, Kubaláková and Kovářová, unpublished). This could be either due to non-quantitative labeling of DNA repeats or due to a random loss of newly synthesized DNA prior to or during flow-cytometric analysis.

Conclusions

PRINS and its variant C-PRINS represent invaluable additions to the repertoire of molecular cytogenetic techniques used in plants. Due to some limitations, which include lower sensitivity and difficulties in simultaneous localization of two different DNA targets, these techniques will probably never replace FISH. Rather, they will be used in particular situations such as rapid localization of highly abundant tandem repeats. Improvement of the PRINSES protocol to achieve quantitative labeling of repetitive DNA sequences on chromosomes in liquid suspension would open new horizons for chromosome sorting in plants. Perhaps the biggest potential of PRINS lies in the ability to discriminate specific repeat variants (sequence subfamilies) of tandemly repeated DNA sequences which can be used to investigate their sequence composition and evolutionary dynamics.

Acknowledgements

This work was supported by research grants 204/04/1207, 521/04/0607, and 521/05/0257 from the Czech Science Foundation, LC06004 and ME 884 from the Ministry of Education,

Youth and Sports of the Czech Republic, and AVOZ50510513 from the Academy of Sciences of the Czech Republic.

References

Abbo, S., Dunford, R. P., Miller, T.E., Reader, S. M., and King, I. P. (1993a) Primer-mediated in situ detection of the B-hordein gene cluster on barley chromosome 1H. *Proceedings of the National Academy of Sciences of USA, 90,* 11821-11824.

Abbo, S., Miller, T. E., and King, I. P. (1993b) Primer induced in situ hybridization to plant chromosomes. *Genome, 36,* 815-817.

Cheng, Z. K., Presting, G. G., Buell, C. R., Wing, R. A., and Jiang, J. (2001) High-resolution pachytene chromosome mapping of bacterial artificial chromosomes anchored by genetic markers reveals the centromere location and the distribution of genetic recombination along chromosome 10 of rice. *Genetics, 157,* 1749-1757.

Cuadrado, A., and Jouve, M. (1997) Distribution of highly repeated DNA sequences in species of the genus *Secale*. *Genome, 40,* 309-317.

Doležel, J., and Lucretti, S. (1995) High-resolution flow karyotyping and chromosome sorting in *Vicia faba* lines with standard and reconstructed karyotypes. *Theoretical and Applied Genetics, 90,* 792-802.

Doležel, J., Kubaláková, M., Bartoš, J., and Macas, J. (2004) Flow cytogenetics and plant genome mapping. *Chromosome Research, 12,* 77-91.

Doležel, J., Kubaláková, M., Bartoš, J., and Macas, J. (2005) Chromosome flow sorting and physical mapping. In K. Meksem, G. Kahl (Eds.), *The Handbook of Plant Genome Mapping. Genetic and Physical Mapping.* (Pp. 151-171) Weinheim: Wiley-VCH Verlag GmbH and Co. KGaA.

Gosden, J., Hanratty, D., Starling, J., Fantes, J., Mitchell, A., and Porteous, D. (1991) Oligonucleotide-primed in situ DNA synthesis (PRINS): a method for chromosome mapping, banding, and investigation of sequence organization. *Cytogenetics and Cell Genetics, 57,* 100-104.

Gualberti, G., Doležel, J., Macas, J., and Lucretti, S. (1996) Preparation of pea (*Pisum sativum* L.) chromosome and nucleus suspensions from single root tips. *Theoretical and Applied Genetics, 92,* 744-751.

Heslop-Harrison, J. S., Murata, M., Ogura, Y., Schwarzacher, T., and Motoyoshi, F. (1999) Polymorphisms and genomic organization of repetitive DNA from centromeric regions of Arabidopsis chromosomes. *Plant Cell, 11,* 31-42.

Koch, J., Kolvraa, S., Petersen, K. B., Gregersen, N., and Bolund, L. (1989) Oligonucleotide-priming methods for the chromosome-specific labelling of alpha satellite DNA in situ. *Chromosoma, 98,* 259-265.

Kovářová, P., Navrátilová, A., Macas, J., and Doležel, J. (2007) Chromosome analysis and sorting in *Vicia sativa* using flow cytometry. *Biologia Plantarum, 51,* 43-48.

Kubaláková, M., Macas, J., and Doležel, J. (1997) Mapping of repeated DNA sequences in plant chromosomes by PRINS and C-PRINS. *Theoretical and Applied Genetics, 94,* 758-763.

Kubaláková, M., and Doležel, J. (1998) Optimization of PRINS and C-PRINS for detection of telomeric sequences in *Vicia faba*. *Biologia Plantarum, 41,* 177-184.

Kubaláková, M., Nouzová, M., Doleželová, M., Macas, J., and Doležel, J. (1998) A combined PRINS-FISH technique for simultaneous localisation of DNA sequences on plant chromosomes. *Biologia Plantarum, 41,* 293-296.

Kubaláková, M., Lysák, M. A., Vrána, J., Šimková, H., Číhalíková, J., and Doležel, J. (2000) Rapid identification and determination of purity of flow-sorted plant chromosomes using C-PRINS. *Cytometry, 41,* 102–108.

Kubaláková, M., Vrána, J., Číhalíková, J., Šimková, H., and Doležel, J. (2002) Flow karyotyping and chromosome sorting in bread wheat (*Triticum aestivum* L.). *Theoretical and Applied Genetics, 104,* 1362-1372.

Leitch, A. R., Schwarzacher, T., Jackson, D., and Leitch, I. J. (1994) *In Situ Hybridization: a practical guide.* Oxford, UK: BIOS Scientific Publishers Ltd.

Lucretti, S., and Doležel, J. (1997) Bivariate flow karyotyping in broad bean (*Vicia faba*). *Cytometry, 28,* 236-242.

Lysák, M. A., Číhalíková, J., Kubaláková, M., Šimková, H., Kunzel, G., and Doležel, J. (1999) Flow karyotyping and sorting of mitotic chromosomes of barley (*Hordeum vulgare* L.). *Chromosome Research, 7,* 431-444.

Macas, J., Doležel, J., Gualberti, G., Pich, U., Schubert, I., and Lucretti, S. (1995) Primer-induced labeling of pea and field bean chromosomes in situ and in suspension. *Biotechniques, 19,* 402-408.

Macas, J., Požárková, D., Navrátilová, A., Nouzová, M., and Neumann, P. (2000) Two new families of tandem repeats isolated from genus *Vicia* using genomic self-priming PCR. *Molecular and General Genetics, 263,* 741-751.

Macas, J., Navrátilová, A., and Mészáros, T. (2003) Sequence subfamilies of satellite repeats related to rDNA intergenic spacer are differentially amplified on *Vicia sativa* chromosomes. *Chromosoma, 112,* 152-158.

Macas, J., Navrátilová, A., Koblížková, A. (2006a) Sequence homogenization and chromosomal localization of VicTR-B satellites differ between closely related *Vicia* species. *Chromosoma, 115,* 437-447.

Macas, J., Navrátilová, A., Kubaláková, M., and Doležel, J. (2006b) PRINS on plant chromosomes. In F. Pellestor (Ed.), *PRINS and In Situ PCR Protocols* (Second Edition, pp. 133-139). Totowa, NJ: Humana Press.

Menke, M., Fuchs, J., and Schubert, I. (1998) A comparison of sequence resolution on plant chromosomes: PRINS versus FISH. *Theoretical and Applied Genetics, 97,* 1314-1320.

Mukai, Y., and Appels, R. (1996). Direct chromosome mapping of plant genes by *in situ* polymerase chain reaction (*in situ* PCR). *Chromosome Research, 4,* 401-403.

Nagaki, K., Cheng, Z., Ouyang, S., Talbert, P. B., Kim, B., Jones, K. M., Henikoff, S., Buell, C. R., and Jiang, J. (2004) Sequencing of a rice centromere uncovers active genes. *Nature Genetics, 36,* 138-145.

Navrátilová, A., Neumann, P., and Macas, J. (2003) Karyotype analysis of four *Vicia* species using in situ hybridization with repetitive sequences. *Annals of Botany, 91,* 921-926.

Nouzová, M., Kubaláková, M., Doleželová, M., Koblížková, A., Neumann, P., Doležel, J., and Macas, J. (1999) Cloning and characterization of new repetitive sequences in field bean (*Vicia faba* L.). *Annals of Botany, 83,* 535-541.

Pich, U., Meister, A, Macas, J., Doležel, J., Lucretti, S., and Schubert, I. (1995) Primed *in situ* labelling facilitates flow sorting of similar sized chromosomes. *Plant Journal, 7,* 1039-1044.

Shi, L., Zhu, T., Morgante, M., Rafalski, J. A., and Keim, P. (1996) Soybean chromosome painting: a strategy for somatic cytogenetics. *Journal of Heredity, 87,* 308-313.

Thomas, H. M., Williams, K., and Harper, J. A. (1996) Labelling telomeres of cereals, grasses and clover by primed in situ DNA labelling. *Chromosome Research, 4,* 182-184.

Vrána, J., Kubaláková, M., Šimková, H., Číhalíková, J., Lysák, M. A., and Doležel, J. (2000) Flow sorting of mitotic chromosomes in common wheat (*Triticum aestivum* L.). *Genetics, 156,* 2033-2041.

Zhu, T., Shi, L., and Keim, P. (1995) Detection by in-situ fluorescence of short, single-copy sequences of chromosomal DNA. *Plant Molecular Biology Reporter, 13,* 270-277.

In: PRINS and PNA Technologies in Chromosomal Investigations ISBN 1-60021-413-4
Editor: Franck Pellestor, pp. 91-96

Chapter 8

In Situ Detection of Chromosomal DNA with Padlock Probe

David Zhang*, Josephine Wu, Fei Ye and Tao Feng
Molecular Pathology Laboratory, Department of Pathology,
Mount Sinai School of Medicine, New York, NY 10029

Abstract

The padlock probe or circularizable oligonucleotide probe (C-probe) is a unique molecule that offers significant advantages over conventional oligonucleotide probes. A closed circular structure forms through ligation of its ends after hybridization onto a target, and locks onto its target due to helical turns formed between complementary sequences of the target and the padlock probe. The nature of padlock and stringent requirement for ligation make this probe especially useful for *in situ* hybridization and detection of single nucleotide polymorphisms (SNPs).

Keywords: Padlock probe, circularizable oligonucleotide, DNA ligation, single nucleotide polymorphism (SNP), in situ detection

Introduction

Oligonucleotide probes provide a useful method for the detection of target nucleic acids by the formation of a double helical structure between complementary sequences based on stringent requirements of Watson-Crick base pairing making hybridization extremely specific. First described by Kool in 1991 [1], the padlock probe or C-probe is a uniquely designed oligonucleotide probe containing three regions: two target complementary

* Correspondence to: David Y. Zhang MD, PhD, MPH Director, Molecular Pathology Laboratory, Mount Sinai School of Medicine, Box 1122, One Gustave Levy Place, New York, New York 10021, U.S.A. Phone: 212-659-8173; Fax: 212-427-2082; E-mail: David.Zhang@MSSM.edu

sequences located at the 5' and 3' termini and an interposed generic linker region [2, 3]. Once the padlock probe hybridizes to a target, its 5' and 3' ends are juxtaposed (Figure 1A) and a closed circular molecule is then generated by incubating the padlock probe-target complex with a DNA ligase. The resulting closed circular molecule is locked on its target due to the helical turns formed between the complementary sequences of the target and the padlock probe [2]. The permanently bound padlock probe allows for stringent washing to remove unbound components, thereby, enhancing assay signal to noise ratios. Therefore, the advantage of the padlock probe is that it offers greater specificity and less background when compared to current conventional probes. Furthermore, once the padlock probe hybridizes to a specific target sequence, it can then serve as a template for linear (rolling circle amplification, RCA) or exponential (ramification amplification method, RAM) amplification and significantly increases the specificity and sensitivity of signal detection.

Recently, two interesting methods were presented for catenation of padlock probes to double-stranded DNA (dsDNA) sequences. Escudé et al. [4] demonstrated the formation of a triplex complex between a dsDNA and a circularizable probe through the interaction between polypurine-polypyrimidine sequences in the dsDNA target and padlock probe. The advantage of the triplex padlock probe is that it allows for binding of the padlock probe to dsDNA without pre-denaturation. Kuhn and colleagues designed two peptide nucleic acid (PNA) probes that are complementary to the opposite strand of padlock probe binding site [5]. These PNA 'openers' can invade the dsDNA and may facilitate binding of the padlock probe to target DNA. Furthermore, they prevent the padlock probe from shifting laterally, resulting in more specific localization of the signal intracellularly.

The padlock probe ranges from 70 to 125 nucleotides long with two target complementary regions each ranging from 15 to 30 nucleotides and a linker region ranging from 30 to 75 nucleotides [2, 3, 6, 7]. The padlock probe can be readily synthesized with most DNA synthesizers. The 5' end of the probe must be phosphorylated and the 3' end must contain a hydroxyl group; both are required by DNA ligase. Addition of a phosphate group to the 5' end is usually achieved by incubation with a polynucleotide kinase in the presence of ATP or during chemical synthesis [8].

Successful ligation of the two ends depends on two factors: both ends must bind adjacent to each other and no mismatch should exist at the ends, especially the first 1-3 nucleotides at its 5' and 3' ends (Figure 1B) [2, 8]. A variety of ligases can be used for this reaction, such as T4 DNA ligase, Taq ligase, Ampligase, T4 RNA ligase [9]. However, ligation efficiency can reach as high as 90% with T4 DNA ligase [5, 8]. Furthermore, Landegren et al. [2] demonstrated the ligation specificity of T4 DNA ligase depended upon increasing the NaCl concentration. While NaCl concentrations between 50 and 150 mM result in optimal ligation activity, at 250 mM NaCl T4 DNA ligase was shown to correctly distinguish a C/G match from notoriously difficult T/G mismatches, resulting in a greater than 1,000-fold difference in ligation rates. Although T4 DNA ligase is efficient for ligation of DNA probe onto a DNA target, ligation of a DNA probe onto a RNA target is very poor [8]. However, the ligation efficiency can reach 80% by addition of $MnCl_2$, low NaCl, low ATP concentration, and high concentration of T4 DNA ligase [8, 10]. The stringent requirement of ligation allows detection of single nucleotide differences between target DNA and probe (Figure 1B).

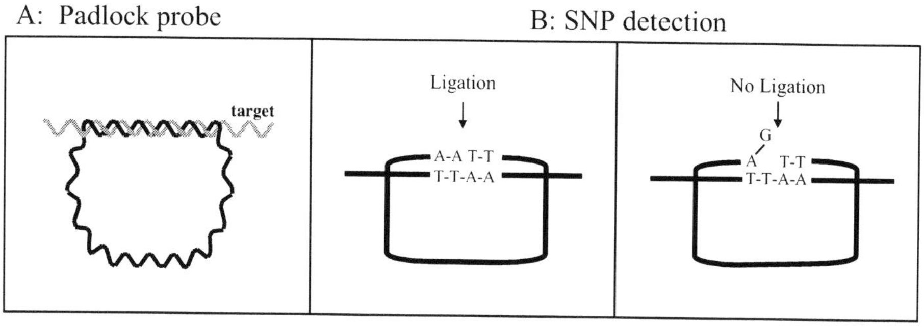

Figure 1. Schematic representation of the padlock probe. (A) A padlock probe hybridizes to its target through its complementary regions and helical turns to form a closed circular probe. This results in the interlocking of the padlock probe onto the target. (B) Only when the ends of the padlock probe perfectly match with its target can ligation occur, thus allowing detection of a single nucleotide polymorphism (SNP)

In Situ Detection of Target DNA

As a result of its high specificity, the padlock probe has found clinical and research applications in several genetic assays, including *in situ* hybridization and SNP detection. The padlock probe was initially designed to analyze chromosomes in a cell undergoing interphase or metaphase [2]. Due to its ability to lock onto target DNA, a glass slide can be washed under stringent conditions (above the melting temperature), therefore lowering background signal and improving detection sensitivity. With the aid of fluorescent reporter molecules, such as biotin or digoxigenin, the bound padlock probes attached to interphase or metaphase chromosomes can be visualized under a microscope [2]. Two fluorescent labeled padlock probes have been used to detect human centromeric alpha satellite DNA of chromosome 13 and 21 which contains multiple repeats [11]. Two chromosomes were individually visualized under a microscope. Furthermore, the padlock probe can detect a single nucleotide difference in two satellite sequences [11]. Although the padlock probe alone is adequate to detect a repeat motif, it is insufficient to detect a single copy gene sequence *in situ* due to low signal intensity.

In order to increase the detection sensitivity, RCA has been applied to amplify the target-bound C-probe [9]. Christian et al. [9] has shown that with RCA, a single copy gene (Tp53) can be detected in HLB cells. Because the RCA products are still bound to the padlock probes which are locked onto the target, the signal remains localized. The authors also found that pretreatment with restriction endonuclease and exonuclease III to remove one of the DNA strand to facilitate the hybridization can significantly improve the detection efficiency from 20-30% to ~90%. The authors further applied RCA to detect several mRNAs (Tp53, p72, vimentin, keratin 10, etc) in a single cell and the level of mRNA expression were quantified with adequate image system.

The C-probe that binds to genomic DNA in a cell can also be amplified with exponential RAM *in situ* [12]. We used RAM technology to detect EBV in both tissue sections and cells from clinical specimens. We have shown that C-probe can hybridize to its target and be ligated and amplified by RAM. This method is more sensitive than conventional hybridization assay.

In situ RCA and RAM are ideal techniques for intracellular target detection. Hybridization of a C-probe to its target and ligation of the two ends results in a covalently "locked on" probe [2] This allows for stringent washings to remove unbound probes, reducing background signals. The large multimeric products (>8,000 nucleotide long) generated during the RCA or RAM prevent amplification products from diffusing out of the cells, a problem inherent to *in situ* PCR [13], thus providing better signal localization. The mild isothermal amplification condition of RAM preserves tissue morphology. Lastly, because the circularizable probe hybridizes only to a small region on the target (50 nucleotides) and because no primer extension on the target nucleic acids is required, cross-linking of the target nucleic acids due to fixation process has a minimal effect on the target-independent aspect of the RCA or RAM reaction [14].

Immuno-RCA, an RCA-based protein detection method, has also been used to detect proteins in routine histological tissue sections [15]. After binding of the primary antibodies to their antigens, the second antibody covalently tagged with a RCA primer along with pre-circularized padlock probe was applied to the tissue section. The RCA products were decorated with biotinylated or FITC-labeled reporter probe and detected by colorimetric deposition. There was a 4-fold increase of signal sensitivity with respect to antibody dilution. The author also demonstrated that immuno-RCA amplified the signal on cell surface which was then subject to flow cytometry analysis. The fluorescent intensity was increased by 30-fold as compared to conventional fluorescent-labeled antibody.

In Situ Detection of Single Nucleotide Polymorphism

SNPs of genomic DNA are considered to be the most common form of genetic variations, occurring in approximately every 250-2,000 bases. With the availability of large SNP databases, detection of specific sets of SNPs will allow for identification of disease-associated polymorphisms to correlate therapies with clinical outcomes and drug effect as well as parentage and forensic testing. The padlock probe offers a significant advantage to detect SNPs. In order for ligation to occur, both ends must perfectly hybridize to the target at juxtaposition (Figure 1B). The discrimination of SNPs by ligation was approximately 1:100 to 1:1,000 in a mixture of mutant alleles in an excess of wild-type alleles [16]. For RNA targets, the discrimination by ligation ranged from 20- to 200- fold [10].

The feasibility of detection single nucleotide difference using padlock probe was demonstrated by Nilsson et al. [11]. The authors were able to detect tow alpha satellite sequences located on chromosomes 13 and 21 which differ in a single nucleotide position. Although each padlock probe carried limited number of signal which can not be visualized individually, many padlock probes bound to the alpha repeats condensed at the centromere of the chromosomes allow fluorophore to be observed under a microscope.

To improve the sensitivity, RCA was applied to amplify the ligated padlock probe [17, 18]. Recently, Larsson et al. [18] described a method to detect single mutation A3243G in mitochondrial DNA. In order to amplify the padlock probe on DNA template and to reduce the non-specific background, the authors introduced several enzymatic steps before amplification. The target double stranded DNA (dsDNA) was initially digested with restriction enzymes to create free ends and a single stranded DNA was created by further digesting the nicked DNA with a 5'-3' exonuclease. The padlock probe was then hybridized to the single stranded DNA and circularized with a DNA ligase. The 3' extra-ssDNA from padlock probe binding site was removed by the 3'-5' exonuclease activity of phi 29 DNA polymerase and the target DNA was used as a primer for RCA, hence, target-primed RCA. With this modification, the authors were able to visualized single mutation in mitochondrial DNA with greater sensitivity and specificity.

Conclusion

The padlock probe represents the next generation of probe-based hybridization with the ability of direct target detection. The padlock probe improves the binding affinity, hybridization stringency and allows more efficient SNP discrimination. With the use of different fluorescent-labeled padlock probes, multiplex padlock assays could greatly increase throughput in genotyping and permit highly sensitive and specific measurement of gene expression. The padlock probe and its amplifications (i.e. rolling circle amplification and ramification amplification) allow for broad applications in clinical testing, research laboratories, and drug discovery. Clinically, it can be used for detection and identification of infectious agents, cancer screening and diagnosis, neonatal genetic diagnosis and parentage, and forensic testing.

References

[1] Kool, E. Molecular recognition by circular oligonucleotides: increasing the selectivity of DNA binding. *J. Am. Chem. Soc.* 1991, 113: 6265-66.

[2] Nilsson, M.; Malmgren, H.; Samiotaki, M.; Kwiatkowski, M.; Chowdhary, B. P.; Landegren, U. Padlock probes: circularizing oligonucleotides for localized DNA detection. *Science* 1994, 265: 2085-88.

[3] Zhang, D. Y.; Brandwein, M.; Hsuih, T. C. H.; Li, H. Amplification of target-specific, ligation-dependent circular probe. *Gene* 1998, 211: 277-85.

[4] Escudé, C.; Garestier, T.; Helene, C. Padlock oligonucleotides for duplex DNA based on sequence-specific triple helix formation. *Proc. Natl Acad. Sci. USA* 1999, 96: 10603-07.

[5] Kuhn, H.; Demidov, V.; Frank-Kamenetskii, M. Rolling-circle amplification under topological constraints. *Nucleic Acids Res.* 2002, 30(2): 574-80.

[6] Kool, E. T. Circular oligonucleotides: new concepts in oligonucleotide design. *Annu. Rev. Biophys. Biomol. Struct.* 1996, 25: 1-28.

[7] Wang, S.;Kool, E. T. Circular RNA oligonucleotides. Synthesis, nucleic acid binding properties, and a comparison with circular DNAs. *Nucleic Acids Res.* 1994, 22: 2326-33.

[8] Hsuih, T. C. H.; Park, Y. N.; Zaretsky, C.; Wu, F.; Tyagi, S.; Kramer, F. R.; Sperling, R.; Zhang, D. Y. Novel, ligation-dependent PCR assay for detection of hepatitis C virus in serum. *J. Clin. Microbiol.* 1996, 34(3): 501-07.

[9] Christian, A. T.; Pattee, M. S.; Attix, C. M.; Reed, B. E.; Sorensen, K. J.; Tucker, J. D. Detection of DNA point mutations and mRNA expression levels by rolling circle amplification in individual cells. *Proc. Natl Acad. Sci. USA* 2001, 98(25): 14238-43.

[10] Nilsson, M.; Barbany , G.; Antson, D.; Gertow, K.; Landegren, U. Enhanced detection and distinction of RNA by enzymatic probe ligation. *Nat. Biotechnol.* 2000, 18(7): 791-93.s

[11] Nilsson, M.; Krejci, K.; Koch, J.; Kwiatkowski, M.; Gustavsson, P.; Landegren, U. Padlock probes reveal single-nucleotide differences, parent of origin and in situ distribution of centromeric sequences in human chromosomes 13 and 21. *Nat. Genet.* 1997, 16(3): 252-5.

[12] Zhang, D.Y.; Zhang, W.; Li, X.; Konomi, Y. Detection of rare targets by isothermal ramification amplification. *Gene* 2001, 274(1-2): 209-16.

[13] Nuovo, G.J.; MacConnell, P.; Gallery, F. Analysis of nonspecific DNA synthesis during in situ PCR and solution-phase PCR. *PCR Methods Appl.* 1994, 4(2): 89-96.

[14] Park, Y.N.; Abi, K.; Li, T.; Hsuih, T.; Thung, S.N.; Zhang, D.Y. Detection of hepatitis C virus RNA using ligation-dependent polymerase chain reaction in formalin-fixed, paraffin-embedded liver tissues. *Am. J. Pathol.* 1996, 149: 1485-91.

[15] Gusev, Y.; Sparkowski, J.; Raghunathan, A.; Ferguson, H.; Montano, J.; Bogdan, N.; Schweitzer, B.; Wiltshire, S.; Kingsmore, S.F.; Maltzman, W.; Wheeler, V. Rolling circle amlpification: a new approach to increase sensitivity for immunohistochemistry and flow cytometry. *Am. J. Pathol.* 2001, 159: 63-9.

[16] Gerry, N. P.; Witowski, N. E.; Day, J.; Hammer, R. P.; Barany, G.; Barany, F. Universal DNA microarray method for multiplex detection of low abundance point mutations. *J. Mol. Biol.* 1999, 292: 251-62.

[17] Zhong, X.B.; Lizardi, P.M.; Huang, X.H.; Bray-Ward, P.L.; Ward, D.C. Visualization of oligonucleotide probes and point mutations in interphase nuclei and DNA fibers using rolling circle DNA amplification. *Proc. Natl. Acad. Sci. USA* 2001, 98 (7): 3940-5.

[18] Larsson, C.; Koch, J.; Nygren, A.; Janssen, G.; Raap, A.K.; Landegren, U.; Nilsson, M. In situ genotyping individual DNA molecules by target-primed rolling-circle amplification of padlock probes. *Nat. Methods* 2004, 1(3): 227-32.

In: PRINS and PNA Technologies in Chromosomal Investigations
Editor: Franck Pellestor, pp. 97-142
ISBN 1-60021-413-4

Chapter 9

Peptide Nucleic Acid (PNA): A Seminal Tool for Biotechnical Applications

Arghya Ray[1*] *and Bengt Nordén*[2*]

[1]Department of Medical Oncology, Dana-Farber Cancer Institute and Department of Medicine, Harvard Medical School, Mayer#534, 44, Binney Street. Boston, MA-02115, USA

[2]Department of Physical Chemistry, Chalmers University of Technology, S 412 96, Gothenburg, Sweden

Abstract

Synthetic molecules capable of binding to gene targets with high specificity are of immense importance because of their potential use as new biomolecular tools for genetic analysis and molecular diagnostics. Out of all synthetic nucleic acid analogs that have been designed in recent years, peptide nucleic acids or PNAs constitute a very special class of analogues, in which the natural sugar-phosphate backbone of nucleic acid has been replaced by a synthetic peptide backbone usually formed from N-(2-amino-ethyl)-glycine units, resulting in an achiral and uncharged mimic. This unique backbone of PNA makes it chemically stable and resistant to hydrolytic (enzymatic) cleavage and it is not expected to be degraded inside a living cell. The molecule is capable of sequence-specific recognition of DNA and RNA and the hybrid complexes exhibit extraordinary thermal stability with unique ionic strength properties. PNA pyrimidines also bind to duplex homopurine-pyrimidine sequences of DNA by a strand invasion mechanism forming a stable PNA-DNA-PNA triplex with a looped-out DNA strand. These quite unique physico-chemical properties of peptide nucleic acids have been drawing major attention at the interface of chemistry and biology/medicine with, potentially, a wide range of applications, including research and diagnostic assays, antigene and antisense therapy and genome mapping. Several PNA-based sensitive methods have been designed

* Correspondence : Prof. Bengt Nordén, E.mail: norden@chembio.chalmers.se;,Phone:(46) (031) 772 3041/3060; Fax:(46) (031) 772 3858;Dr. Arghya Ray; E.mail:_arghya_ray@yahoo.com; arghya_ray@dfci.harvard.edu; Phone:001-617-632-4312; Fax: 001-617-632-6638.

over the past few years encompassing a variety of applications including polymerase chain reactions, detection of genomic polymorphisms and mutations, cytogenetics for rapid *in situ* chromosomal investigation. In this review we shall discuss some recent advancements in the field of PNA science and the usefulness of PNA-based applications in the field of biology and medicine.

Key Words: PNA, DNA, PNA-DNA hybridization, Antigene, antisense, gene therapy, cytogenetics, molecular diagnostics, Nucleic acid biosensor

Abbreviations and Acronyms

DNA	Deoxyribonucleic Acid
RNA	Ribonucleic Acid
PNA	Peptide Nucleic Acid
HPLC	High Performance Liquid Chromatography
(MALDI-TOF)	Matrix Assisted Laser Desorption Ionization Time-of-Flight Mass Spectrometry
PCR	Polymerase Chain Reaction
PARC	PNA Assisted Rare Cleavage
CE	Capillary Electrophoresis
CGE	Capillary Gel Electrophoresis
BIA	Biomolecular Interaction Analysis
SPR	Surface Plasmon Resonance
QCM	Quartz Crystal Microbalance
UV	Ultra - violet
Tm	Thermal Melting Temperature
CD	Circular Dichroism
LD	Linear Dichroism
NMR	Nuclear Magnetic Resonance
SNP	Single-Nucleotide Polymorphism
MD	Molecular Dynamics

Introduction

Peptide nucleic acids or PNAs (Figure 1) are DNA analogs where the normal phosphodiester backbone of DNA has been replaced by a 2-aminoethyl-glycine linkage [1,2]. The nucleotide bases, both natural and in some cases artificial, are connected to the backbone at amino nitrogens via a methyl carbonyl linker. This novel synthetic molecule has quite interesting physico-chemical properties. First of all, PNAs are achiral, non-ionic and not degraded by hydrolytic (enzymatic) cleavage. They are capable of sequence specific binding to DNA and RNA counter parts [3,4] and the PNA-nucleic acid complexes exhibit extraordinary thermal stability with unique ionic strength properties.

Initially, this nucleic acid analogue was considered as a potential drug candidate for anti-gene and anti-sense therapy. However, today's *'PNA-science'*, i.e., the applications of PNA-based technology in the area of biotechnology and medicine [5,6,7,8,9, 9a] is not limited to the area of gene therapy, but also has impact on fundamental research of nucleic acids and the development of diagnostics, including cytogenetics, industrial microbiology, environmental analysis and clinical tests. This review will highlight recent advancements of PNA-based applications in biology and medicine. We will discuss its mode of action in relation to potential use as antigene/antisense agents [10,11,12] and as a novel tool for genetic/cytogenetic analysis and nucleic acid manipulation [13,14,15,16]. We will also deal with the issue of the cellular delivery of peptide nucleic acids in the light of recent progress in this area. Finally, recent improvement of the basic architechture of PNA [17,18,19,20], the subsequent generations of PNA molecules and their applications will be discussed.

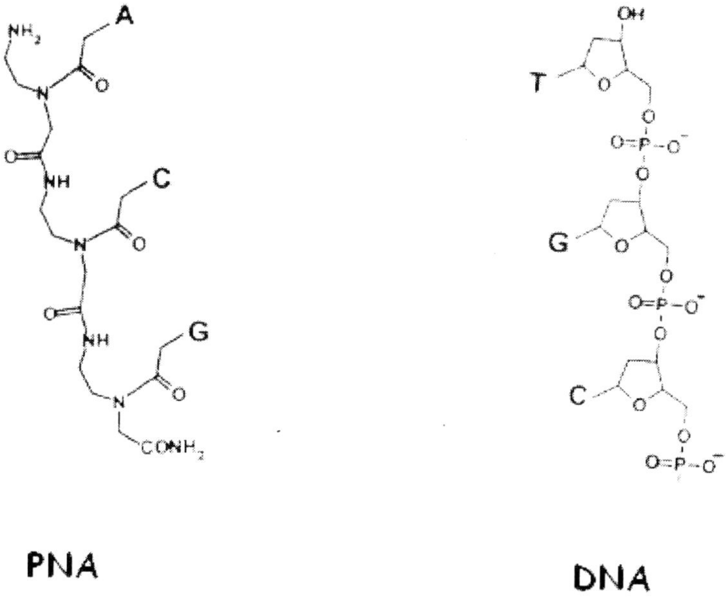

Figure 1. Chemical structure of PNA (left) and DNA (right)

Physico-Chemical Properties

In this section we have included, in brief, the primary aspects of synthesis and physico-chemical properties of PNA. We believe that this will provide the readers with a better understanding of different interesting biological properties of PNA.

Synthesis and Purification

In peptide nucleic acids, 2-aminoethyl-glycine linkages replace the regular phosphodiester backbone of DNA [1,2] and the nucleotide bases are connected to this backbone at the amino nitrogens through a methylene carbonyl linker (Figure 1). Since PNA is an achiral molecule, it can be synthesized without the need of any stereoselective pathway. The PNA oligomers are generally prepared using standard solid phase synthesis protocols for peptides following either Boc or Fmoc chemistry [21-27]. The bis-PNAs are prepared in a continuous synthesis process by connecting two PNA segments via a flexible linker composed of multiple units of either 8-amino-3,6-dioxaoctanoic acid or 6-aminohexanoic acid [28]. It appears that by changing cytosine to a pseudoisocytosine residue in the Hoogsteen strand abolishes the pH dependence of a DNA•PNA-DNA triplex without affecting the stability of the triplex [28]. The subsequent generations of PNAs involve unusual modifications of the backbone or a chimeric architecture. In a PNA/DNA chimera, the PNA and DNA oligomers are fused together to give rise to a chimeric architecture to increase the solubility of the molecule [29]. Attempts have also been made to conjugate a wide range of effector molecules with PNA to increase the solubility and cell permeability. Among these, of particular interest are small peptide fragments, derived from functional proteins, and capable of rendering their specific properties to the conjugate. In recent years, much attention has been on the development (synthesis) of PNA –peptide conjugates with specific properties based on both solid phase and fragment ligation (For detailed discussion see Ref [30]).

Earlier reports by Koch et. al., [31] and Betts et al. [32] described the synthesis of a PNA pentadecamer carrying a heptapeptide and PNA-peptide-PNA chimera, respectively, using traditional PNA chemistry. The presence of a peptide moiety does not significantly affect the structure of the PNA-DNA duplex in either of the cases. Moreover, Betts et al.'s construct is also capable of forming a stable triple-helical complex with an oligopurine complex [32]. The recent developments of PNA-peptide conjugate synthesis with interesting properties have been reported by Nielsen, Corey's and Pandey's groups [33-35]. Recently, the synthesis of a novel polydiamidopropanoate dendrimer PNA-peptide chimera used for the non-invasive magnetic resonance imaging of cancer has been reported by Amirkhanov and Wickstrom [36]. We will discuss PNA-peptides conjugates in terms of their biological properties in later sections in details.

As a part of the postsynthesis procedures, the PNA oligomers are cleaved from the solid support by certain chemical treatments, and purified by HPLC [23,28,37, 38]. Further purification and subsequent characterization are normally made using reverse-phase HPLC followed by matrix-assisted laser desorption ionization time-of-flight (MALDI-TOF) mass spectrometry (23,28, 37, 38).

Physicochemical Properties

Different PNAs and PNA-DNA complexes have been well characterized over the years by a combination of various spectroscopic, crystallographic and computational methods.

In general, the concentration of a PNA oligomer is determined by monitoring the absorbance at 260 nm and at 80 °C; considering that at this temperature complete destacking of the nucleobase monomers should occur and contribution from an ordered backbone is negligible [39,40]. The chemical stability of PNA is quite different from its DNA counterpart.

In general, PNA is fairly acid stable. However, under some basic conditions, a cleavage of the N-terminal PNA unit by ring closure can take place leading to some chemical instability [41]. The charge neutral PNA oligomers have generally poor water solubility and tend to aggregate, to a degree dependent on the sequence of the oligomer. The limited solubility is also related to the length of the oligomer and its purine:pyrimidine ratio [42]. The water solubility is improved to a great extent with the incorporation of charged residues in the oligomer. It could be either positively charged residues (lysine as a C-terminal or backbone modification in place of glycine) or negatively charged ones (modifications, especially for PNA-DNA chimeras) residues that enhance the water solubility of the polymer. Recent developments of several PNA conjugates, especially the PNA-peptide conjugates seem to have taken care of the solubility problem as each of these individual modifications tend to impart greater solubility [30]. Riguet and coworkers have shown that PNA conjugated to neamine, a core moiety of the aminoglycoside antibiotic neomycin B, renders higher solubility and confers a unique RNA cleavage property to the conjugated molecule [43].

The peptide nucleic acids can hybridize to complementary nucleic acid, both DNA and RNA sequences in a sequence dependent manner. Formation of multi-stranded structures containing both Watson-Crick and Hoogsteen base-pairing is also possible from a PNA-DNA interaction, especially triplex structures where PNA and DNA maintain a 2:1 ratio (Figure 2) [44-47,47b,47c]. Since PNA hybridizes to DNA in a highly sequence-dependent manner, the hybridization is strongly affected by base mismatches compared to the corresponding DNA-DNA hybridiation. In fact, PNA oligomers have been shown to maintain sequence discrimination up to the level of a single base mismatch.

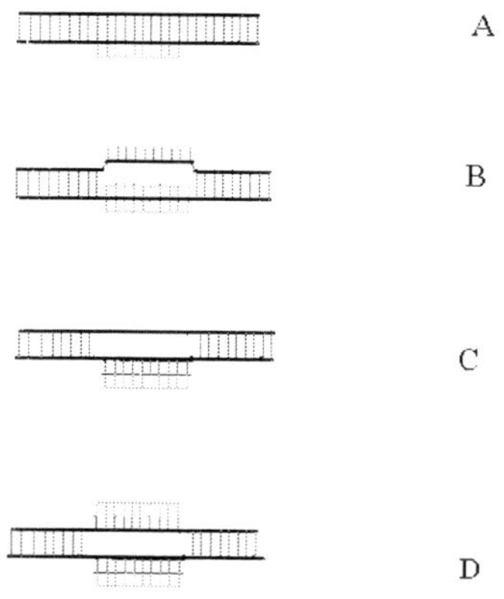

Figure 2. Schematic representation of different molecular complexes formed upon target specific binding of PNA to DNA. (A) Formation of a prototype triple helical structure only possible with cytosine rich homopyrimidine PNAs. (B) Stable triplex invasion complex formed by homopyrimidine PNAs binding to a homopurine DNA target. (C) Duplex-invasion complex (D) Double-duplex invasion complex only possible with PNAs containing nonstandard nucleobases.

PNA-DNA/RNA and PNA-PNA Duplexes

A PNA oligomer, with complementary mixed-sequence DNA, forms a right-handed double helix; as well as with RNA [3]. This observation of Egholm et. al. [3] also defines a "polarity" of PNA, i.e., the direction is denoted as "antiparallel" because of its higher thermal stability, in analogy with the antiparallel structure of normal duplex DNA. It is important to point out here that this is just a convention without physical basis. Thus, a PNA oligomer can form an anti-parallel duplex with complementary DNA sequence, which will have a higher thermal transition or "melting" temperature (Tm value) than the corresponding DNA-DNA complex. Similarly, a parallel PNA-DNA duplex is also possible to form with a thermal transition temperature lower than that of the anti-parallel complex with the same PNA [4]. In contrast to this fact, the thermal stability of a PNA-RNA duplex can be higher compared to a corresponding PNA-DNA duplex [48]. Also, PNA-DNA hybrids are more or less unaffected by changes in ionic strength. As PNA binds to the complementary DNA in a sequence-specific manner, any sequence mismatch that is present in the duplex will severely affect the thermal stability of the PNA-DNA duplex by lowering its thermal transition temperature. Similar all-DNA hybrids will respond to the base-mismatch in a less stringent way as elaborated briefly in Table 1 (for details see ref. [9]). The PNA-DNA chimeras generally follow the Watson-Crick hybridization scheme, and the thermal melting temperature of duplex formed between a chimera and its DNA counterpart usually falls between that of the corresponding PNA-DNA and DNA-DNA duplexes. It is worth mentioning that this value also depends on the PNA and DNA in the chimera strand and the nature and position of the linker molecule between PNA and DNA bases [29]. Esposito et al. [49] reported synthesis and characterization of PNA-DNA chimeras forming quadruplex structures and showed that the UV melting profile of their synthesized chimera is quite similar to the reference PNA structure. Similar biophysical and biochemical properties of PNA-DNA chimeras have also been reported by Bajor et. al [50].

Table 1. Thermal stabilities of PNA-DNA duplexes

PNA sequence	DNA sequence	Melting temperature
H-egl-GGCAGTGCCTCACAA-NH_2[a]	5′-TTGTGAGGCACTGCC-3′[b]	72.3 °C
	5′-TTGTGAG*A*CACTGCC-3′	58.1 °C
	5′-TTGTGAG*A*CA*T*TGCC-3′	39.0 °C
H-egl-GGCAGCGCCTCACAA-NH_2	5′-TTGTGAGGCGCTGCC-3′	> 85 °C
	5′-TTGTGAGGC*A*CTGCC-3′	69.9 °C

[a] Self-melting of this PNA sequence is 44.7°C

[b] The thermal melting temperature of the full complementary DNA duplex (5-TTGTGAGGCACTGCC-3` and its complementary sequence) is 42.2 °C

A stable PNA-PNA duplex can also be formed from the binding of a PNA oligomer to its complementary oligomer. Wittung et al. [51], for the first time, demonstrated the formation of a PNA-PNA double helix; an entirely artificial construct. Because of the achiral nature of the monomeric units of PNA, but preferred helical structure when forming a duplex, PNA first hybridizes into a racemic mixture of right and left handed helices, which then, by "helical seeding", inverts to the helicity preferred due to the presence of a single chiral amino-acid residue at the end (further investigated in Ref. [52]).

For example, the PNA oligomer H-GTAGATCACT-NH2 can form both parallel and antiparallel duplexes with complementary PNA and DNA oligomers and the thermal transition temperatures of this repertoire of structures are as follows: anti-parallel PNA-PNA > PNA-DNA > DNA-DNA are 67 ^{O}C, 51 ^{O}C and 33.5 ^{O}C, respectively, and the parallel PNA-PNA has a melting temperature of ~47 ^{O}C. The parallel PNA-PNA duplex is more stable than the corresponding DNA-DNA duplex [51]. The absence of any significant electrostatic repulsion between the interactive strands is the underlying cause of the much higher thermal stability of a PNA-PNA duplex relative to the corresponding DNA-DNA duplex. PNA-PNA-PNA triplex constructs are also possible from a combination of one purine and two pyrimidine PNA decamers [20,52,53, 53b]. The base-pair geometry of anti-parallel right-handed PNA-containing helices is not much different from that of a B- or an A-form DNA helix. However, the parallel PNA-DNA and PNA-RNA duplexes have quite different kinds of base stacking [20,53-57]. An earlier report from our group had also shown that the incorporation of a PNA strand (or strands in a triplex) would reduce the binding of minor groove binders and completely inhibit binding by intercalation [58]. The cationic ligands do not interact with PNA-DNA double helices or PNA-DNA-PNA triplex in the same way as with the pure DNA counterparts, which is a result of the non-ionic character of the PNA backbone and hence supreme stacking stability. Consequently, while minor groove binders interact with both the duplex and triplex PNA-DNA helices, no intercalator has been found that can acquire enough electrostatic and hydrophobic binding energy to compensate the cost of destacking the DNA-PNA helix (or PNA-DNA-PNA triple helix) and therefore show no binding [58].

The thermodynamic aspects of PNA binding to DNA have been studied by several groups over the years [39, 59-69]. As shown by Tomac et al. [39], the ionic-strength dependence of a PNA-DNA duplex is remarkable and quite different from that of DNA-DNA duplex.. While DNA-DNA is stabilized by increasing NaCl concentration in the range 1 mM to 100 mM, PNA-PNA shows a decreased melting temperature. Interestingly however, in the regime of very high salt concentration (more than 1 M) both DNA-DNA and PNA-DNA duplexes show a similar decreased stability as a result of dispersive interaction, the so-called Hofmeister-Ninham effect, an effect that follows the Hofmeister series as function of choice of anion of the salt.

Certain PNA samples could create problems with certain thermodynamic measurements owing to their low solubility [50,60]. Studies of thermodynamic parameters [39,59,60,62,66] have given important mechanistic insight about the stability of PNA as well as DNA (for practical procedures, see [60,62]). The higher stability of a PNA-DNA duplex over a DNA-DNA duplex is due to a favourable entropic contribution and the higher stability of an all PNA duplex over an all DNA duplex is attributed mainly to the enthalpic contribution (39,59). However, the enthalpic changes ΔH are quite similar for PNA-DNA and DNA-DNA duplexes. Since PNA has a charge neutral backbone, at low ionic strength conditions, a favorable enthalpy of hybridization is observed, generally, in this order: PNA-PNA>PNA-

DNA>DNA-DNA. It is important to note that hydrophobic interactions play an important role for the stabilization of PNA-containing duplexes. In the presence of terminal lysine residues, the change of enthalpy and entropy are different from that determined for charge neutral PNAs without terminal lysine residues. The thermal stability of PNA duplexes, pure as well as hybrid, are strongly dependent on the sequence and secondary structure and so is the thermodynamic information derived from these data [61,65,66,69]. In PNA·DNA duplexes, the thermal stabilization is proportional to the number of G·C base pair content. The stabilization also increases when purine bases are present in the PNA strand but almost not in the DNA strand. For the same set of base-stackings, the backbone influence/chemistry, i.e. aminoethyl glycine for PNA versus deoxyribose for DNA, has tremendous influence on the thermal stability of the hybrid duplexes and related thermodynamic parameters. It has been shown recently by Nielsen's group that in case of PNA-DNA duplexes, having a purine-rich PNA strand will render more stability (~6 kcal/mol) over a pyrimidine rich PNA strand, although the stacked base pairs are similar in both the cases. Moreover, in iso-sequential PNA·DNA duplexes, in the presence of organic solvent, a significant increase in binding enthalpy and entropy is noticed for hybrid duplexes containing pyrimidine-rich PNA strand [69].

Table 2. Synthetic DNA fragments of the cystic fibrosis gene and corresponding PNA probe [14]

Nucleic acid fragment	Sequence	Mutation
Wild type (WT)	5′---TGGCACCATTAAAGAAA-ATATCATCTTTGGTG-TTTCCTATGATGAATATA----3′	None
ΔF508	5′---TGGCACCATTAAAGAAA-ATATCAT---TGGTG-TTTCCTATGATGAATATA----3′	Deletion of three bases
MU1	5′---TGGCACCATTAAAGAAA-ATATC**G**TCTTTGGTG-TTTCCTATGATGAATATA----3′	One-base substitution
MU2	5′---TGGCACCATTAAAGAAA-ATAT**AC**TCTTTGGTG-TTTCCTATGATGAATATA----3′	Two-base substitution
PNA (15 mer)	H-CACCAAAGATGATAT-lys-NH_2	----------

PNA/DNA Triplexes

Theoretically, PNA/DNA triplexes can be of two forms; either PNA_2/DNA or PNA/DNA_2. The former structure is possible when two homopyrimidine PNAs bind to complementary DNA sequences (Tm > 70 °C). A PNA/$(DNA)_2$ triplex is formed from the interaction between a C-rich PNA and GC rich DNA duplex. The cytosine involved here to form the base triplet is protonated. Hence, the resultant triplex structure is undoubtedly pH-dependent. The incorporation of pseudo-isocytosines ("J bases") in place of normal cytosine can make the triplex structure independent of pH [9,28]. From the early works on triple-helical structures involving PNAs, it had been demonstrated that PNA-T_8 forms right-handed PNA_2/DNA triple helix with poly(dA), similar to a standard poly(dA)-(poly(dT))$_2$ triplex [70]. This in fact, was the very first detection of the formation of a PNA-DNA-PNA (pyrimidine-purine-pyrimidine) triple helix [70]. A (PNA)$_2$-RNA triple helical structure is also possible with a thermal transition temperature value similar to that of its (PNA)$_2$-DNA counterpart [71]

In general, PNAs can form triple helical structures upon binding to target double-stranded DNAs by four different binding modes (Figure 2). They are referred as standard triplex formation, duplex invasion, triplex invasion and double duplex invasion. The first three modes of binding would require homopurine/homopyrimidine DNA targets. In triplex and triplex invasion complexes, a homopyridine PNA oligomer binds to homopurine target site in double-stranded DNA. The triplex invasion complexes are much more stable than conventional triplex structures. The duplex invasion complexes are formed with some homopurine peptide nucleic acids and are considerable unstable compared to the previous two structures. A double-duplex invasion complex, whereby two PNA strands can simultaneously bind (invade) the two DNA strands, is very stable but would require non-standard nucleobases on the PNA strands: pseudocomplementary PNA oligomers containing 2,4-diaminopurine and 4-thiouracil instead of adenine and thymine, respectively. Each of these strand invasions is favored by the presence of internal AT rich regions (at least 50%) in the target duplex DNA and by moderately low ionic strength (< 50mM NaCl, best performed with 10 mM EDTA-containing buffers) [72]. In a recent study by Frank-Kamenetskii and coworkers, it has been found that for PNA strand invasion occurring in a typical range of salt concentration (~10-20 mM NaCl), the strand-invading PNA binding rates do not depend on the presence of non-target DNA in the reaction mixture. If the salt concentration is less than 10 mM NaCl, the presence of unrelated DNAs will slow down the rate of PNA binding (or invasion) to its targets [73].

From the early studies of the crystal structures of the PNA_2-DNA triplex it was revealed that the purine DNA strand, while bound to the homopyrimidine PNA in an anti-parallel orientation by Watson-Crick base pairing, makes a second set of base-pairing hydrogen bonds (Hoogsteen) to the second PNA strand of identical sequence running parallel with the DNA: the backbones of these two strands make another set of hydrogen bonds with each other [32]. The hybrid PNA-DNA-PNA triplex has a characteristic helicity, which is perhaps more close to the A-form helix, but never to the B-form. This is not just due to the replacement of sugar-phosphate backbone by PNA backbones but most likely due to distinct conformational preferences of the PNA strands [20,32, 54-57,74-76]. Recent studies of both conventional and a subsequent generation of PNAs have elucidated the structural aspects of peptide nucleic acids in great detail including the handedness of PNA structure [77-84]. The thermodynamics of the melting of PNA_2/DNA, formed from a single-stranded DNA containing an A10 target and bis-PNA consisting of two T10 decamers, indicates that the absolute values of the changes of enthalpy and entropy (values) are two to three times higher than the corresponding variations of thermodynamic parameters for decameric PNA/DNA duplexes of various nucleic base sequences [85], which is the origin of the extremely high stability of the triplexes.

Apart from duplex and triplex invasions of PNA, a quadruplex invasion complex is also possible as was reported earlier by Datta and Armitage [86]. The same group had also reported the formation of a PNA_2/DNA_2 hybrid complex resulting form the binding of homologous PNA sequences to a dimeric G-quadruplex telomeric sequence d(G(4)T(4)G(4)). The hybrid G-quartet has preference for Na+ over Li+ [87]. The quadruplex structure is becoming more and more important in the field of *PNA science* with the discovery of telomere recognition of PNA and formation of quadruplex complex by PNA/DNA chimeric structures [88]. Moreover, guanine-rich PNA oligomers alone can also form quadruplex-like structures [89, 90]. A recent thermodynamic study of the PNA-DNA quadruplex-forming chimeras has revealed a higher thermal and thermodynamic stabilities for chimeric quadruplexes compared to the corresponding unmodified DNA quadruplex and demonstrated the presence of kinetically stable intermediate species in the quadruplex formation process [91].

PNA - An Important Tool in Molecular Biology

The tremendous potential of peptide nucleic acids as efficient tool for a wide range of molecular biology applications has been established over the years. We will review some of the applications here.

Use of Peptide Nucleic Acids to Modulate Polymerase Chain Reaction (PCR)

Peptide nucleic acids on their own cannot directly interact with DNA polymerase. However, they could affect the rate of a PCR reaction by their mere presence in the reaction mixture, either by binding to the template or competing with the DNA primers. Small PNA oligomers may block the template so that the latter becomes unavailable for intra- and inter-strand interaction during reassociation. PNA-DNA chimeras, on the other hand, are recognized by DNA polymerase and can be used as primers for DNA polymerase-driven reactions [92,93]. The fact that the special non-canonical phosphate backbone-architecture of PNA-DNA chimeras is recognized by the DNA polymerase is quite intriguing and it suggests that the primer phosphate backbone may not be an essential factor for the polymerase recognition.

The specificity of PNA-DNA interaction and the higher stability of hybrid duplex compared to a DNA duplex have also been utilized in the detection of genetic mutations by PCR. The novel technique, which has been utilized by many groups over the years, is termed as PNA-directed PCR clamping (Figure 3). In this method, strategically, PNA is targeted against one of the PCR primer sites and the PNA oligomers compete with the conventional PCR primers. Since this step is carried out at a temperature higher than that for conventional PCR primer annealing, the PNA/DNA complex formed at one of the primer sites after selective binding effectively blocks the formation of a PCR product.

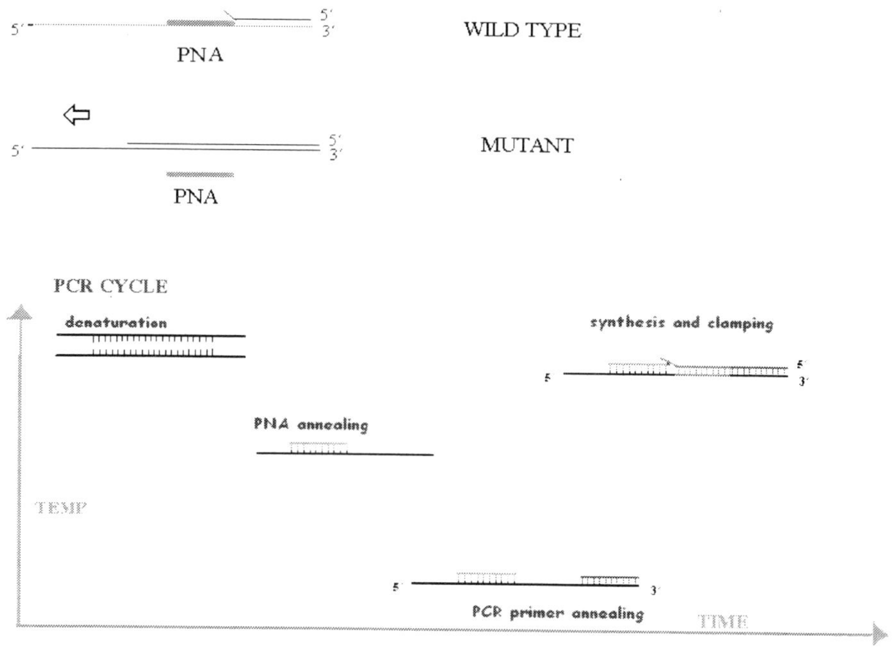

Figure 3. Mutation analysis using PNA directed PCR clamping: Schematic representation of the strategy for the PCR cycle involving PNA directed clamping. In case of the normal (wild type) DNA, the bound PNA will sterically hinder annealing of a partially overlapping primer sequence thus preventing the normal sequence from appropriate PCR amplification. In case of mutant alleles, the melting temperature of the PNA/DNA is reduced and the primer can outcompete PNA annealing to carry on preferential amplification of mutant sequences.

This powerful strategy was first reported by Örum et al. [13, for review see Ref. 94] to demonstrate that this type of PNA clamping could discriminate three different point mutations at a single position. The method has been used for sensitive detection of mutations related to neoplastic diseases [15, 95,96,97], factor V genotyping [98], amplification of functional immunoglobulin light chain [99], detection of apolipoprotein B mRNA editing [100] and mitochondrial DNA mutations [101]. Fiandaca et al. 2001 [102] developed a novel fluorogenic method of PCR based on the use of PNA. The method is called Q-PNA PCR (quencher-labeled peptide nucleic acid polymerase chain reaction) and utilizes a quencher-labeled PNA (Q-PNA) directed against a 5' tag sequence of a fluorophore–labeled DNA oligo primer. The primer fluorescence is quenched by the PNA-DNA hybridization. As the PCR reaction proceeds, the quencher PNA or Q-PNA is displaced due to incorporation of the primer into duplex amplicons. This will cause the fluorescence of the labeled-dye to increase proportionately with increasing concentration of amplicons in the reaction mixture. The Q-PNA method was also utilized by Alsmadi et al. [103] in a rolling circle amplification based assay for increased detection sensitivity.

Of later date is the PNA directed PCR-clamping method, which may be used quite rigorously, for the clinical diagnosis of malignancies related to mutation of various oncogenes. A diagnostic biochip array uses PCR-clamped probe for mutant-specific amplification and detection of mutation in codon 12 of K-ras proto-oncogene [104]. The preexistence of imatinib mesylate-resistant clones in chronic myelogenous (CML) leukemia was detected by a very sensitive PNA-based PCR clamping technique [105]. Through a combination of PNA clamping and fluorescence hybridization, the detection sensitivity for the Bcr/Abl mutations Thr315Ile, Glu255Lys, and Tyr253His was effectively enhanced in such a way that 1 mutant cDNA molecule could be detected in 500 negatives (0.2%) [105]. Horny et al. [106] reported the detection of a c-kit point mutation Asp816-Val from the bone marrow infiltrates of systemic mastocytosis (SM) patients by PNA mediated PCR clamping [106]. This method has also been extended to detect Ki-Ras in tissue and plasma samples of patients with pancreatic cancer [107] and different activating mutations of the GNAS gene in peripheral blood cells in McCune-Albright syndrome [108].

In recent days, this clamping method has also been used in microbiological and food sample analysis. By using PNA-mediated PCR clamping Takiya et al. have developed methods for rapid identification of Escherichia coli O157:H7 [109] and detection of non-hotspot mutations in the Ames test using Salmonella typhimurium TA98 [110]. Using PNA-derived approaches, the first method identifies a single nucleotide (T93G) alteration in the uidA gene, a mutation specific to E. coli O157 strain, while the second method describes a cost-benefit attempt to rapidly select non-hotspot mutants among hisD+ revertants of Salmonella typhimurium TA98 in Ames test [109,110]. Moreover, the clamping assay has also been used for the partial evaluation of genetically modified organism content in food [111].

PNA Hybridization as Alternative to Southern Hybridization and Bioanalytical Applications Based on PNA Hybridization

Southern hybridization is a commonly used technique in molecular biology, but there are certain disadvantages of this process as it is laborious and may sometimes yield poor sequence discrimination between closely related species. PNA pre-gel hybridization has emerged as a simplified alternative to Southern hybridization. In this process, fluoresceinated (labeled) PNA-oligo probes are subjected to hybridization to a denatured DNA sample at low ionic strength. The reaction mixture is run on a gel to separate single-stranded DNA fragments on the basis of their length. The bound complex will have higher mobility compared to the excess unbound PNA. These hybrid duplexes are thereafter transferred to a nylon membrane, dried, UV crosslinked and detected using standard chemiluminescent techniques [7]. Igloi has reported several methods of automated detection of point mutations by electrophoresis using PNA-containing gels [112,113,114; reviewed in 115]. The method commonly known as affinity electrophoresis, requires the entrapment of PNA probe the gel matrix thereby affecting (retardation) the mobility of complementary oligonucleotides. This allows sequence-specific hybridization during electrophoretic passage of a complementary DNA, which also makes it sensitive to mismatch [112,113]. A similar sensitive gel-mobility shift assay to detect PNA oligomers was also reported by Jensen and Richelson [116].

Instead of using a standard electorphoretic method for separation, the bound peptide nucleic acids can also be separated by using capillary electrophoresis (CE). The strong affinity PNA-DNA hybridization combined with capillary electrophoresis follows a very simple and straightforward protocol, and has given rise to a powerful bioanalytical diagnostic method for specific sequence detection with simultaneous size separation of the target DNA. Based on this strategy, several protocols for screening genetic mutations have been reported over the years. We will review some of them here.

In capillary electrophoresis, the separation process is fast and can be visualized on-line by either ultraviolet absorbance or laser-induced fluorescence (LIF) detection systems. Although, it is possible to analyze minute amounts of sample (of the order of pico- or femtograms) in conventional capillary electrophoresis, perhaps, the major disadvantage is that it is not possible to carry out the simultaneous analysis of multiple samples (for details see ref 9 and references therein).

A novel PNA-based method for the detection of genetic mutation using *free solution capillary electrophoresis* was first reported by. Carlsson et al. [14]. The method was demonstrated to be extremely sensitive and by using 15-mer PNA probes, it could detect a single-mismatch in a 50-mer sample DNA representing a part of the cystic fibrosis transmembrane conductance regulator (CFTR) gene (Figure 4). The separation is carried out at high temperature (~ 70° C and ionic strength ~50 mM), at which the PNA-wild type DNA duplex will remain stable but hybrid complex carrying single–mismatch DNA will be melted. The free excess PNA will bind to the capillary wall and will not be detected. The PNA-bound fully matched DNA is detected using *free solution capillary electrophoresis* [14]. The same technique was also used to detect point mutation in a 15-mer DNA fragment representing a vulnerable part of the p53 gene [Nordén and co-workers, 1998; unpublished data]; the mismatch represented here corresponds to a mutation in the codon 175, where an arginine is replaced by a histidine in the altered P53 protein. A combination of CE and LIF was also used by Basile et al [117] to assess the use of PNA probes for detecting DNA single-base mutations. In a model system based on cystic fibrosis R553 and R1162 single-base mutations, an efficient CE separation of the PNA/DNA hybrids from both double-stranded and single-stranded DNA was achieved up to a resolution of single base pair discrimination. [117]. The capillary electrophoresis method was also applied for direct chiral separation of chiral peptide nucleic acid (PNA) monomers using charged cyclodextrins as chiral selectors [118]. The PNA probes bearing chiral monomers based on D-lysine have been used in CE for the detection of cystic fibrosis R553X point mutation. In this way, it was possible to discriminate among mutated homozygous, normal homozygous and heterozygous subjects indicating a clear advancement towards the diagnosis of cystic fibrosis, as well as for other genetic mutations [119].

The RNA-binding properties of PNAs have also been characterized utilizing capillary gel electrophoresis (CGE) [120], which compared to free solution capillary electrophoresis, shows a stronger dependence on mass and does not always follow a 1:1 stoichiometry like PNA-DNA duplexes [120]. The sequence-specific binding of DNA to liposomes containing di-alkyl peptide nucleic acid amphiphiles has been characterized from capillary zone electrophoresis (CZE) [121]. Moreover, the CE method was also used to evaluate the intracellular biological activity of PNA after conjugation with a cell-penetrating synthetic model peptide [122]. The peptide-conjugated PNA would have 3-fold and 8-fold higher intracellular concentrations in neonatal rat cardiomyocytes and CHO cells, respectively, compared to the concentrations after exposure of the cells to PNA not conjugated to signal peptides.

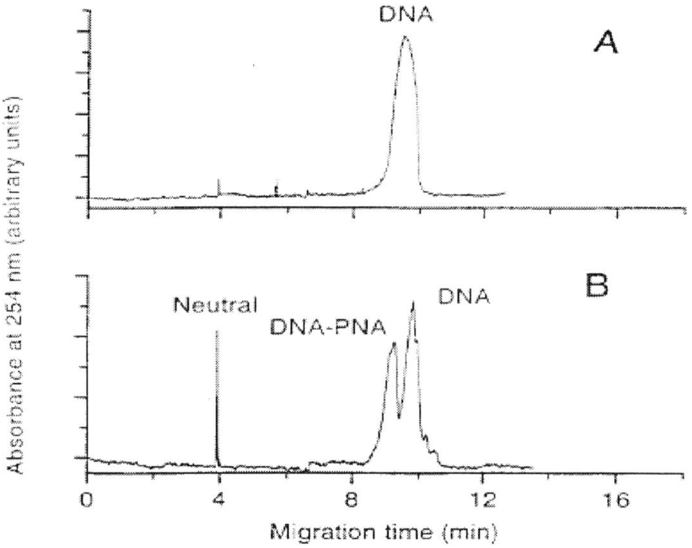

Figure 4. Free-solution capillary electrophoresis of a mixture of PNA and DNA. Panel (A) represents the electrophoregram at 70°C of a mixture of PNA and DNA (MU1, Table 2). The latter has a single-base substitution in the recognition region. At 70°C, PNA can not remain bound to the mutant DNA sequence and the peak corresponding to PNA-DNA hybrid duplex will not appear. Panel (B) is the control electropherogram. at the same temperature representing the interaction of the same PNA (Table 2) with the wild type DNA sequence. (Figure adapted by permission from Carlsson et al., 1996).

The unique hybridisation properties of PNA have led to the development of technologies over the years to using this novel probe for sequence-specific biosensors. The PNA hybridization and corresponding mismatch analysis can be studied using a BIAcore (*B*iomolecular *I*nteraction *A*nalysis) instrument by evaluating a real-time biomolecular interaction analysis between PNA probes attached to the surface of the sensor chip and in solution analyte partners, either DNA or PNA. The detection principle in BIA utilizes surface plasmon resonance (SPR) [for details of the instrumentation principle see ref 123].

The principle was utilized for the first time by Nordén and co-workers [48] to study the association and dissociation kinetics of PNA-DNA/RNA hybridisation. An interesting finding in this study is the increased association rates of matched compared to mismatched sequences. It clearly indicates the importance of "probing pre-equilibria" in the process of recognition, as compared to a case where run-on rate is purely determined by diffusion. The PNA probe could be either biotinylated, attached to dextran-streptavidin coated gold surface of the sensor chip [48], or cysteinylated, immobilized employing the strong coupling between gold and sulfur (Nordén et. al., 1998; unpublished observations). In the latter case, short spacer molecules like mercaptohexanol, are also used together with the cystein-PNA probe to form the monolayer at the top of the sensor (gold) surface in order to prevent DNA from being non-specifically adsorbed to the surface. Several other groups have subsequently utilized the SPR and BIAcore principles to study PNA-based biosensor applications [124-128].

Several other PNA-based mircoarrays and biosensors have been reported [129] including the application of quartz crystal microbalance [130, 131], MALDI-TOF mass spectrometry [37,38,132-135] and Potentiometric applications [16]. A detailed PNA-based microarray and biosensor applications has been reviewed by Brandt and Hoheisel [136]. Moreover, a genetic screening based on the colorimetric variations of cyanine dye binding to DNA-PNA hybrid has been proposed as a general method of genetic screening [137].

Antigene and Antisense Applications of PNA

Peptide nucleic acids are potential candidates for development as gene therapeutic drugs. In principle, gene therapeutic drugs can be adapted to work in two different ways (Figure 5). Peptide nucleic acids are designed to recognize and bind to complementary sequences in a particular gene. This binding interferes with the transcription of that particular gene (antigene strategy).

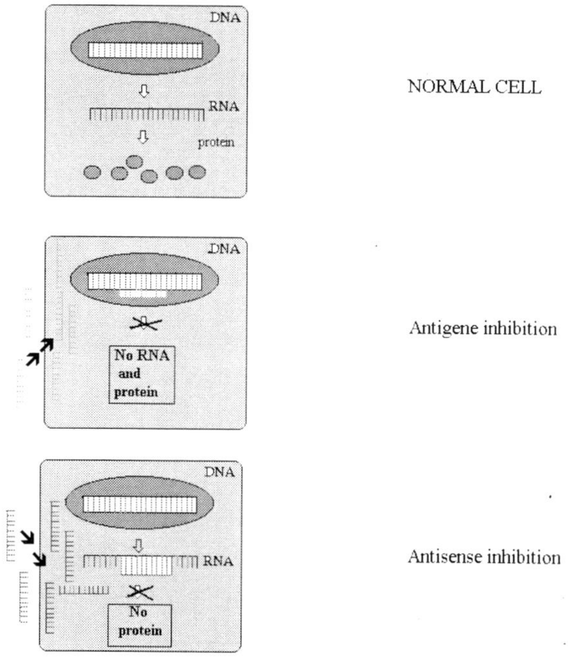

Figure 5. Antigene and antisense strategy: An *antigene* oligomer (e.g., PNA) could bind to a complementary sequence in the DNA and inhibit transcription of the gene. On the other hand, cells can also be treated with an *antisense* oligomer and hybridization to a specific mRNA sequence can inhibit the expression of a protein at the level of translation

Alternatively, there can be antisense PNAs directed against complementary sequences in mRNA to inhibit its translation (antisense strategy). Chemically and biologically stable PNAs have shown potential to significantly affect replication, transcription and translation processes without any sign of general toxicity, as revealed from *in vitro* studies. Here, in the following sections we will discuss all three aspects of PNA therapeutics and highlight some of the recent advancements in the field of PNA directed gene therapy.

Inhibition of Transcription

Peptide nucleic acids are capable of arresting transcription by forming stable triplex/strand-invaded/strand-displacement complexes with target DNA sequences. Such stable complexes, formed either at the promoter region or located further downstream, could block the function of both prokaryotic and eucaryotic RNA polymerase and thus are capable of working as antigene agents. The stable complex formed at the promoter region restricts further DNA access of the polymerase. On the other hand, PNA strand-displacement complexes, generally located at the downstream away from the promoter, can block polymerase progression and transcription elongation resulting in the formation of truncated RNA transcripts; the poly-purine DNA target (of PNA) must be present in the gene of interest. It was demonstrated in earlier studies that PNA oligomers as short as 8-mers (e.g., PNA-T_8) are capable of blocking phage T3 polymerase activity [138] and small PNAs could be directed against the promoter region to shut down the activity of the gene of interest as shown for the alpha chain of interleukin-2 receptor (IL-2Rα) gene [139]. However, in those cases the PNA strand invasion or the formation of strand displacement complex at physiological salt concentration was quite slow [42,140], which required certain modifications of PNA for the purpose of an improved PNA-DNA binding at physiological conditions. It was found that modifications like chemically linking the ends of the Watson-Crick and Hoogsteen PNA strands to each other [28], introducing pH-independent pseudoisocytosines into the Hoogsteen strand [28] or positively charged lysine residues [28,141], incorporating intercalators [142], in PNA strand can increase the association rates with dsDNA. Under physiological conditions, PNA binds to supercoiled plasmid DNA at a much faster rate compared to its binding to linear DNA [140,141]. Since transcriptionally active chromosomal DNA is usually negatively supercoiled and the binding of PNA to dsDNA is enhanced when the DNA is being transcribed, this can serve as a better target for PNA *in vivo*. Moreover, the PNA binding occurs with increased efficiency when the PNA target is situated on the non-template strand instead of the template strand. The transcription-mediated binding of peptide nucleic acid (PNA) to double-stranded DNA results in the repression of the RNA polymerase activity, the latter is a consequence of its own action and hence referred *suicide transcription* [142]. It is also interesting to note that a looped-out single-stranded structure formed during a PNA strand invasion can also act as initiation site for *E. coli* and mammalian RNA polymerases whereby the polymerase might start transcription using the single-stranded loop as a template [139,144].

Moreover, PNA/DNA chimeras and PNA-peptide conjugates with improved cellular delivery can further facilitate the antigene (as well as antisense) activity of PNA. Several new generation PNAs directed against different critical as well as clinically relevant targets in various genomes have shown strong therapeutic potential. These include interrupting multiple steps in the replicon of HIV-1 [145] as well as targeting genes responsible for different kind of malignancies and other diseases [146-155].

Inhibition of Translation

Peptide nucleic acids can deliver the antisense effects either by exerting a steric blockage on the translation machinery due to the formation of PNA-RNA complex or a ribonuclease H (RNase H)- mediated cleavage of the RNA strand in PNA/DNA-chimera-RNA hetero-duplex [29,154]. PNA/RNA duplexes, on its own cannot act as substrates for RNase H and the antisense effect is due to the steric blocking of either RNA processing, transport into cytoplasm or translation. Early reports on the anti-sense propertied of PAN concluded that both duplex- (mixed sequence) and triplex-forming (pyrimidine-rich) PNAs could inhibit translation at targets overlapping the AUG start codon [71]. Triplex-forming PNA (TFP) can block the translation machinery at targets in the coding region of mRNA, however, the translation elongation arrest requires a $(PNA)_2$-RNA triplex and thus homopurine tract of 10-15 bases is necessary at the target site. TFPs can inhibit translation and ribosome elongation at initiation codon and codon region targets, respectively. Duplex-forming PNAs are unable to produce this effect.

Antisense PNA oligomers, when directed against different parts of the same target gene at the same time can show additive to synergic antisense effects [156]. A combination of three PNAs directed against three different parts of PML/RARα gene could efficiently inhibit translation at a low concentration much below the critical concentration used for each individual. Bis-PNA or clamp-PNA structures are capable of forming internal triple helical PNA_2.RNA constructs and if they are targeted against the coding region of mRNA, can also cause a stop in translation [71].

Antisense PNAs can specifically inhibit bacterial gene expression. However, bacteria cannot adapt to RNA-interference mechanism, as, in general, they do not have the system. The conventional antisense strategy requires efficient delivery of PNA in the cell. Thus modifications, that aid cell permeation, are likely to increase the antisense effects of PNA. Early works by Good and Nielsen with 'leaky' mutant strains of *E. coli* [157] demonstrated that antisense effects of PNA targeted against β-galactosidase and β̃lactamase genes were indeed capable of down-regulating the expression of these two genes [157]. Along the same line, it was shown that bis-PNAs, targeted against the homo purine stretches in rRNA, could inhibit the ribosome function in a cell-free system [158] and would cause translation arrest at a submicromolar range of PNA concentrations. Moreover, AS19 *E. coli* strain can also be inhibited by the same set of PNAs at low micromolar concentration [158]. Subsequent studies by Good and coworkers probed chemical modifications to improve the efficacy of standard PNA. When conjugated with a cell membrane-active KFFKFFKFFK peptide, 9- to 12-mer PNA oligomers render improvements in antisense potency in E. coli without affecting the target specificity [159]. Similar peptide modification has been shown effective against different genes of *Staphylococcus aureus* at a low micromolar concentration [160]. In a recent study by the same group, an attempt was made to target peptide-PNA conjugate against the essential gene inhA of *Mycobacterium smegmatis* . This resulted in growth inhibition and caused an alteration of the cell morphology of the subject similar to the conditions in InhA-depleted cells [161]. Several others groups have also tried different modifications of PNA to improve its antisense activity. These modifications favor improved PNA uptake in the cell with improved antisense activity. A 9-aminoacridine-PNA conjugate has been shown to be effective in delivering PNA to human cancer cells and help down-regulation of MDM2 and activation of p53 gene [162]. It is also possible to down-regulate genes by using negatively charged and highly soluble PNA analogs with alternating phosphonates (HypNA-pPNAs) and this method has been shown to be effective in a zebra fish model [163]. In addition to the modifications described above, PNA conjugated to different cell penetrating peptides enhance the cellular uptake of PNA and hence its antisense activity. PNAs conjugated to a cell-penetrating peptide, transportan, enhance their cellular uptake and this has been shown to down-regulate the expression of GalR1 protein in human melanoma cell line [164]. Corey and co-workers showed that antisense PNA conjugated to either peptide D(AAKK)(4) or to another peptide containing a nuclear localization sequence (NLS) could be easily delivered into cultured cells.

One important finding is that, although DNA/lipid-mediated delivery requires less PNA, peptide-conjugated delivery of PNA is simpler less toxic to primary cells [165]. Moreover, PNAs can also be conjugated to cationic peptides like tat and oligoarginine for an improved delivery to the cells via an endosomal pathway. The presence of calcium ions can enhance the antisense effect of peptide nucleic acids by promoting the endosomal release [166]. Recently, it has been discovered that just a simple cationic peptide motif like octa(l-lysine) peptide, covalently linked to PNA, can promote the cellular uptake of the antisense agents into BCL1 cells as well as primary murine macrophages [167].

Other strategies that have been explored recently for efficient delivery of antisense PNAs include PNA adsorption onto polymeric microspheres and encapsulation in autologous erythrocytes. Chiarantini et al [168] demonstrated that antisense activity of unmodified PNAs is possible even at nanomolar concentration following both of these delivery mechanisms. One of the most recent advancement in this area was the development of short antisense chimeric PNA conjugated to a polypeptide containing eight Ser-Arg repeats $(SR)_{(8)}$. This modification, known as PNA-SR, has been shown to be able to modulate the splicing of bcl-x gene both *in vitro* and *in vivo* [169]. Antisense PNA has also been applied in radioimaging by using different types of PNA conjugates. A biotinylated antisense PNA was attached to a streptavidin-coated monoclonal antibody (MAb) conjugated to the transferrin receptor (TfR). The PNA-TfR-MAb when administered intravenously in mice was transported across the transmembrane, from blood to the cytosol of brain cells. PNA, labeled with $^{(111)}$In, was targeted against 20-37 nucleotides of the rat glial fibrillary acidic protein (GFAP) mRNA; the gene is downregulated 24 hrs after reversible brain ischemia with a middle cerebral artery occlusion (MCAO). Antisense PNA hybridizing to the target mRNA sequence and decreased GFAP gene expression in cerebral ischemia was examined and imaged *in vivo* [170]

Inhibition of Replication

PNA-mediated inhibition of replication is possible by the blockage of DNA-polymerase-catalyzed elongation of DNA primers. DNA replication can also be inhibited by PNA-induced strand invasion of DNA duplex under physiological conditions, or by the formation of any single-stranded DNA segment during the replication process. The fact that extra-chromosomal mitochondrial DNA is mostly single-stranded during replication, makes it a good target for PNA. This type of inhibition of the replication of human mitochondrial DNA creates a new avenue for the treatment of patients suffering from ailments related to the heteroplasmy of mitochondrial DNA. Since both wild type and mutated DNA are present in the same mitochondria, experiments could be designed to selectively inhibit the replication of mutated DNA under physiological conditions without affecting the wild type DNA [171]. Inhibition of the mutated population allows selective propagation of the wild-type population and thereby shifts the turnover [172]. Several modifications have been proposed to improve the mitochondrial import and attachment of PNA to the target sequence. PNAs conjugated to lipophilic phosphonium cations (phosphonium-PNA or ph-PNA or benzophenone-PNA derivatives carrying a triphenylphosphonium moiety) can be easily imported to mitochondria through the lipid bilayer driven by the membrane potential across the inner membrane. This type of mitochondrial localization of PNA probes by the addition of a caged lipophilic cation would result in irreversible binding selectivity between wild type and mutated mtDNA that differ only by a single nucleotide [173,174].

There have other reports of inhibition of replication, either by site-directed inhibition of DNA replication by triple helix formation [175] or by direct inhibition of plasmid replication [176]. Another useful application in this direction is the inhibition of human immunodeficiency virus-1 (HIV-1) replication by PNA and PNA-conjugates [177-179]. PNAs directed against HIV-1 genome can inhibit both TAR and Tat-dependent viral replication. The replication and gene expression of HIV-1 is highly dependent on the specific interaction between the viral protein Tat and its trans-activation responsive element (TAR). The latter forms a highly stable stem-loop structure and is thus a potential target of PNA-directed anti-HIV-1 activity [35,177, 178]. Several PNA modifications for an improved anti-HIV-1 activity have been reported. Modifications like cyclic PNA [182], PNA-neamine conjugate [43], PNA linked to a nuclear localization signal (NLS) [183], and PNA-peptide conjugates have been shown to substantially suppress HIV-1 replication [184-187]. PNAs conjugated with different membrane transducing peptides (MTD) transportan, penetratin, tat peptide, transportan-27 and two truncated transportan-27 derivatives, transportan-21 and transportan-22 can be targeted to the TAR region of the HIV-1 genome [184] as all of them are inhibitory to HIV-1 replication *in vitro*. Moreover, these conjugates are also potent virucidal agents, which render HIV-1 virions non-infectious upon brief exposure. Moreover, cell penetrating peptides like transportan or a novel chimeric peptide R6-Penetratin, can inhibit Tat-dependent trans-activation in a HeLa cell assay in a dose-dependent manner [186, 187].

Applications of PNA in Molecular Cytogenetics

There has been a recent interest to utilize peptide nucleic acid-directed *fluorescence in situ hybridization* (PNA-FISH) assays as tools for rapid cytogenetics analyses. The high-affinity sequence-specific binding properties of PNAs make them better probes over the conventional long DNA probes. There are certain advantages of using the PNA-FISH technique. First of all the, highly specific binding of PNA will generate lower background signal, an important criterion of the cytogenetics. Moreover, PNA oligomers with charge neutral backbone can bind to target genomes at a very low ionic condition, not suitable for the binding of complementary sequences of the target molecules. This prevents the re-annealing of complementary strands and in particular, disfavors the renaturation of repeat sequences. The PNA probes are extremely stable and compatible with standard reporter fluorophores with low photo bleaching.

Historically, the PNA-FISH technique was first used for telomere analysis [188], in which the use of fluoresceine-labeled PNAs allowed precise quantitative estimation of individual and global telomere length in metaphase chromosomes. PNA probes targeted to the telomeric repeats in human metaphase chromosomes rendered better staining of telomeres compared to standard oligo probes and could distinguish between fluorescence signals coming from each individual sister chromatid ends. Following this, there have been reports involving PNA-probes directed against telomeres; both in situ PNA-FISH and other studies [151, 189,190, 191,192,193]. Two basic methods, namely *Q-FISH* and *FLOW-FISH*, have been developed over the years to study telomere length regulation in cells (Figure 6). Q-FISH method relies on appropriate PNA probes and a digital image analysis systems needed to quantify the fluorescence signals. With a resolution in the range of ~200 bp, this method can determine the intra-chromosomal distribution of telomere length (p-arms vs. q-arms), individual length of all telomeres in a given genome, accurate average length of all telomeres, and identify the shortest telomere and its location. [for review see ref. 189]. Two images are taken; one of metaphase chromosome spread and another the image of the dye-bound telomere, stained with DAPI and red dye/Cy3/5, respectively. The data stored by the image acquisition program are then analyzed by combining these two images using fluorescence image analysis software. Flow-FISH, on the other hand, is a fast-track modification of the Q-FISH method where the high resolution of the latter is compromised for a rapid detection of telomeric fluorescence signal and measurement of average length of telomere. Instead of metaphase chromosomes, interphase cells are used in this procedure and measurements are by flow cytometry. Although the overall analysis can be done very fast, it is not possible to analyze the telomere length in individual chromosomes and hence, ideally, Q- and FLOW-FISH should be done in parallel with each other [189]. Well-defined algorithms are already in place for a detailed image/data analysis and the results of which are in excellent agreement with the conventional telomere repeat fragment Southern blot method [193]. The fact that each cell has an inherited pattern of telomere length, which is normally defined in the zygotic stage, was also shown by PNA-FISH studies [194]. Whatever is imprinted in zygote is maintained throughout the life and thus identical homologous telomeres from two monozygotic (aged) twins would show less difference in relative telomere length than two homologues within one individual.

The method has also been tested against different transformed cell lines (of human or non-human origin) to measure the extent of change of telomeric length before and after treatment with different mutagens. These results could be extrapolated to actual clinical setups and used as first- hand 'data-tool' for the treatment of patients with anti-cancer agents [195-197]. From recent studies, the PNA-FISH protocols for the estimation of telomere length have been optimized in chronic myelogenous leukaemia (K-562), multiple myeloma (IM 9), T cell lymphoblastic leukaemia (1301) and T-cell acute lymphoblastic leukemia (CCRF-CEM) cell lines [198, 199].

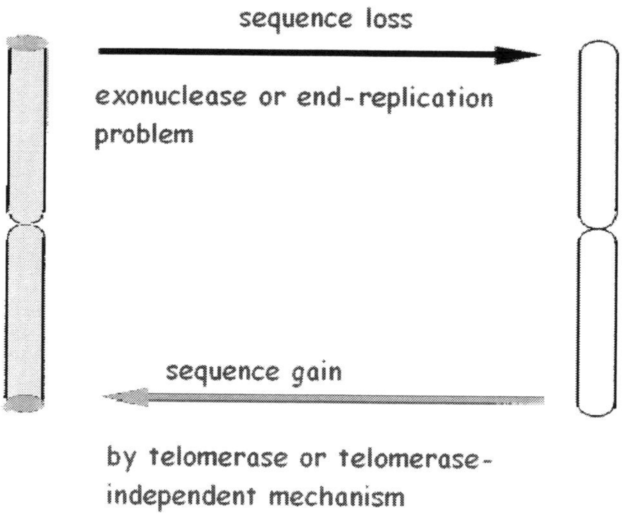

Figure 6. Schematic representation of the mechanism of telomere length maintenance (adapted from Ref. [189])

In human T-cells, a stable hierarchical order exists in terms of the chromosomal telomere length profile, as detected by PNA-FISH, which is genetically determined and conserved between different human cells [200]. Moreover, the action of mutagens like bleomycin and mitomycin C and specificity of telomerase PNA-inhibitors have also been studied [201,202]. Lymphocyte cultures, when exposed to mutagens like bleomycin and mitomycin C, are affected which is manifested by an alteration in telomere length. Telomere-specific PNA-FISH probes are able to detect the changes in treated cells [201]. It was concluded that effects of mitomycin C on the telomeric repeat area are more than that for bleomycin. The PNA-FISH probes revealed that at mitomycin C concentration of 1 ug/ml, significant shortening of telomeres of the chromosome arms; 2q, 3p, 5q, 7p, 10q, 11p, 13q, 17p, 18pandq, and 21q would occur. On the other hand, at 8 ug/ml concentration of bleomycin, the telomeres at chromosome arms, 1q, 6p, 17p, 20pandq, and 22q are only minimally affected in terms of the signal from the fluorescence-PNA probes [201]. Some recent advancement on the existing protocols combines PNA-FISH methods with other protocols to give rise to a new methodology [203,204]. Flow-FISH-DDD (dye dilution and DNA staining), is one of such which combines flowcytometry with dye dilution and DNA staining (with DAPI) to measure telomere-specific fluorescence in proliferating cells [203]. With the advancement of technology, it is now possible to do IQ-FISH (Quantitative-FISH on interphase nuclei) to accurately quantify the telomere length in interphase nuclei [204]. With this new PNA-based method, morphological details of a larger number of cell populations can be obtained.

The current PNA-FISH application is not restricted to telomore characterization and over the years progress has been made to apply specific PNA probes for the other kinds of chromosomal characterization and clinical investigation, including the *in situ* detection of abnormalities [for review see 151, 205-209]. The sequence specific binding properties of PNA, combined with *in situ* fluorescence, was used to analyze the spatial distribution of CTG trinucleotide repeats in human myotonic dystrophy (DM) and visualize repeat expansion within the DM gene located on chromosome 19 [210]. Since, peptide nucleic acid-FISH probes retain the sequence (single base) discrimination property of the parent (non-labeled) PNA oligomers, it is possible to distinguish between two centromeric DNA repeat sequences that differ by a single base pair, as shown by Chen et al for Centromere Protein B (CENP-B) repeats located on mammalian chromosomes [211]. Detection of chromosomal aberrations has been possible by the use of different chromosome-(allele-)-specific PNA probes, which has advanced the basic understanding of genetic polymorphism and its clinical relevance. PNA probes specific for chromosomes 1, 2 and 4 have been used to detect complete and incomplete chromosomal exchanges in human lymphocytes irradiated with low dose-rate tritium beta rays [212]. Short (~14-20 mers) PNA oligomers directed against alpha-satellite domains, each specific against one particular chromosome, were used on metaphase and interphase nuclei and specific probes against chromosomes 1, 2, 7, 9, 11, 17, 18, X, and Y were identified [213]. A simple multicolor FISH method using directly labeled 18-22 mer PNA probes for chromosomes 1, X, and Y was also tested in interphase and metaphase cells from normal human lymphocytes and fibroblasts with abnormal chromosome contents [214].

Pellestor and coworkers have improvised this PNA technology to working with human spermatozoa and provided a new direction for *in situ* aneuploidy estimation [215]. Successful application of PNA probes to this kind of complex biological samples is of particular interest due to its tremendous potential in the area of in situ cytogenetics. The efficiency of PNA labeling on human sperm was validated from comparative estimates of disomies X, Y, and 1 using sperm preparations from donor subjects. Multicolor FISH (standard), primed *in situ* (PRINS), and PNA procedures were run in parallel and all three resulted in similar *in situ* nuclear labeling disomy rates. The hybridization kinetics of PNAs on sperm was comparable to the kinetics of the PRINS reaction, and both were way faster (less than 1 hour) than the same for standard FISH technique (overnight hybridization) [215]. This PNA-based methodology has been further extended by the same group to perform studies on isolated human oocytes, polar bodies and blastomeres [216] using specific PNA probes for each of the chromosomes 1, 4, 9, 16, 18, X, and Y. All together a total of 34 *in vitro* unfertilized oocytes and 23 blastomeres were tested by using both simple and sequential multicolor PNA labeling, and the results were in good agreement with the same from previous FISH studies. This particular application is simple and has tremendous potential for preimplantation genetic diagnosis. This and other new approaches would help overcoming the earlier technical limitations related to *in situ* analysis of human meiosis, reassure the importance of cytogenetic investigation of human oocytes and create a new horizon in the understanding of aneuploidy processes [217]. Although multiple DNA probes are still successfully used, as shown by Codina-Pascual et. al. [218] in a FISH characterization of a dicentric Yq (p11.32) isochromosome in an azoospermic male, novel PNA based approaches have become the method of choice for better efficiency compared to DNA-based studies [205,217]. In a recent study by Agerholm et al. [219] on sequential FISH analysis in human blastomeres, three sets of labeled PNA probes were used in a competitive displacement method. The first set of probes was all labeled, but the second and the third sets consisted of both labeled and unlabeled probes. The key here is the unlabeled probe, which corresponded to the labeled probe used in the previous cycles. This resulted in a destabilization of already bound probe at 50°C before the next FISH cycle and thus up to five sequential FISH cycles could be performed before the loss of any signal. PNA probes specific for chromosome 1, 13, 16, 17, 18, 21, X and Y were used and the study showed that several FISH cycles can be carried out in this way without compromising the signal as well as the fidelity [219].

We would like to mention here in brief about another important application of fluorescence in situ hybridization technique using PNA probes. PNA-FISH has emerged over the years as an important tool for rapid microbiological analysis and diagnosis of infectious diseases. Peptide nucleic acids have been used as probes in fluorescence *in situ* hybridization assay to differentiate between tuberculous and nontuberculous mycobacterium species in smears of mycobacterium cultures and detect Mycobacterium tuberculosis in smear positive sputum samples [220-222]. The detection of gram-negative and gram-positive bacteria was possible using species-specific PNA probes as has been reported against Escherichia coli, Pseudomonas aeruginosa, Staphyloccocus aureus and Salmonella [223,224]. It has been shown that identification of Staphylococcus aureus from positive blood culture is possible by using Staphylococcus aureus PNA FISH [225-228]. Moreover PNA-FISH has also been proposed for use in microfluidic devices [229] to detect food and water borne pathogens Campylobacter jejuni, Campylobacter coli and Campylobacter lari. [230] and as a probe to differentiate between closely related species like Sorex araneus and Sorex granaries [231]. Another recently reported application of PNA hybridization was the use of PNA as a probe in piezoelectric gene sensor microarray for direct detection of hepatitis B virus (HBV) genomic DNA [232].

Other Interesting Molecular Biology Applications

Several other interesting PNA-assisted molecular biology applications have been developed over the years. Although, the techniques have not been extensively applied by that many groups, however, their merit lie on the smart application of PNA properties to develop these novel methodologies. In combination with methylases and other restriction endonucleases, peptide nucleic acids can act as rare genome cutters [8]. The method, known as PNA Assisted Rare Cleavage (PARC) (Figure 7) technique employs the strong sequence-selective binding of PNAs, preferably bis-PNAs, to short homopyrimidine sites on large DNA molecules [for details, see ref 8, reviewed in 9]. PNAs can be used in an artificial restriction enzyme system [5] to purify target nucleic acids utilizing nickel affinity chromatography [6]. The interaction of PNA with different enzymes has also been reported. PNAs cannot directly stimulate RNase H activity after duplex formation with RNA.

However, chimeric DNA/PNA structures have been shown to stimulate the RNase H activity by forming the chimeric RNA double strand [29]. Moreover, it has also been demonstrated that a short internal phosphodiester window within a PNA structure is sufficient enough to activate RNaseH-dependent cleavage (Figure 8) of a targeted RNA [233]. Similarly, direct or indirect interaction of PNA with DNA polymerase, reverse transcriptase and Telomerase have been reported [10,234-238].

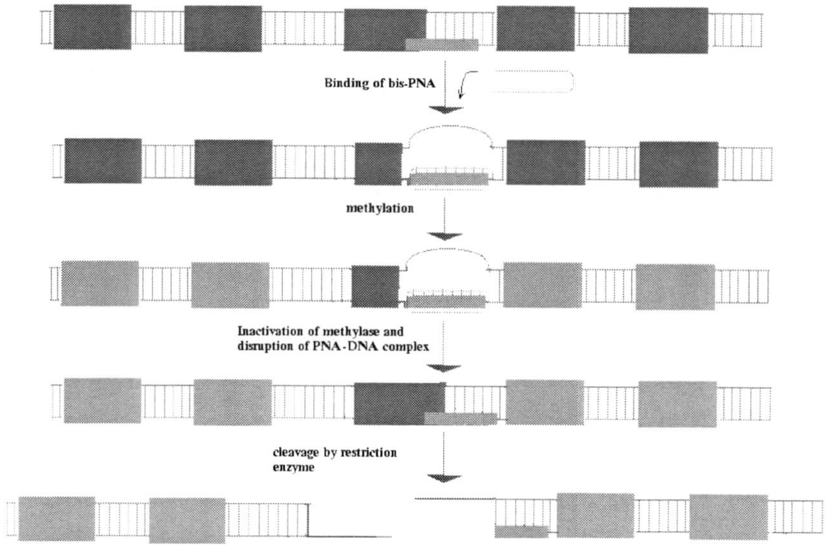

Figure 7. The PARC technique: Symbols (—), (■), (■) represent PNA binding site, restriction enzyme site and methylated restriction enzyme site, respectively. Methylated sites are not subjected to a restriction digestion

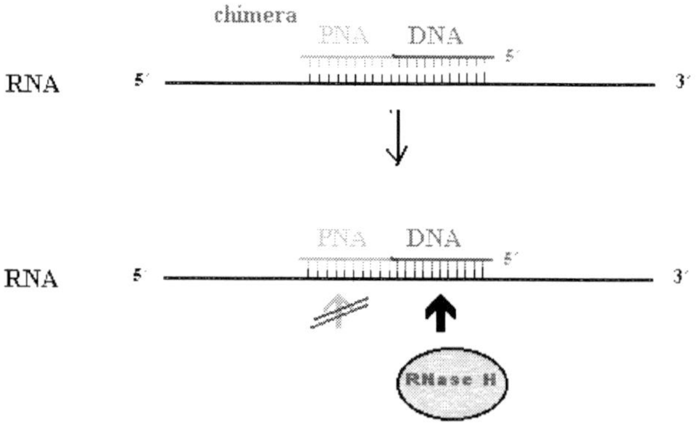

Figure 8. Schematic representation of RNase H mediated cleavage activity following the binding of a PNA-DNA chimera to an RNA target

Cellular Delivery of PNA and *In Vivo* Activity

The cellular uptake of peptide nucleic acid is very poor, a challenge that must be overcome before it could be used as a therapeutic drug. In earlier works regarding the cellular activities of PNA, techniques were used to bypass the membrane barrier. Several modifications have been applied to improve the cellular activity of PNA and its potency as a therapeutic drug.

Cellular effects of PNAs with antisense activity against a transfected gene have been demonstrated by cellular microinjection of PNA oligomers [10, 18]. Modification of PNA molecules, either by the modification of base-peptide units or conjugating suitable potential ligand molecules to it, enhance the possibility of physical or receptor-mediated cellular uptake [239-242]. Introduction of 'guide' sequences or 'vector' peptides to PNA [240-242] has been helpful as these modifications help in proper docking of PNA oligomers to the cell membrane. There had been some early attempts to increase cellular permeabilization of PNA in eukaryotic cells using streptolysin O [239], lysolectin [12] and Tween [237]. Noteworthy to mention there are some reports about direct PNA uptake in different eukaryotic cells, e.g., rat embryo fibroblasts [243] and human myoblasts [171], at very high concentrations (~20 μM). In comparison, PNA-DNA chimeras are taken up by cells at a relatively lower extracellular concentration (~1 μM) and considered better antisense agents, as reported for Vero or NIH3T3 cells [244,245]. The kinetics of chimera uptake is similar to what is expected for an oligonucleotide with identical sequence. However, there is a possibility of cytotoxic effect at high concentration of PNA.

Currently, the modifications that are being undertaken to increase the cellular uptake of PNA can be divided into three broad categories. Two of them involve simple conjugation of another non-PNA moiety, either a peptide or charged lipid, to a PNA molecule resulting in favourable cellular internalisation of the conjugated molecule. A third approach is to use a vehicle/vesicles, e.g., liposomes, to transfer PNA into the cellular system.

Peptide nucleic acids, coupled to a retro-inverso delivery peptide, have been shown to be rapidly internalized (within a minute) in cultured neurons [246]. The peptide construct used in the study, similar to the concept that have been used for the synthesis of retro-inverso PNA [247,248], was a novel retro-inverso analog of the 16-mer p-Antp peptide, and was coupled to an antisense PNA against the AUG translation initiation region of prepro-oxytocin mRNA. Although the penetration of non-conjugated single PNA molecules was very slow compared to the conjugated one, both the constructs were able to lower the amounts of mRNA coding for prepro-oxytocin in these neurons [246]. Similar retro-inverso peptide was also used for better PNA uptake in Raji lymphoma cells [249]. Peptides have been used to enhance the delivery of antisense oligomers as noncovalent complexes as well as covalent conjugates [250, also see ref 122]. A hexameric or nanomeric PNA oligo could be coupled to a transport peptide (transportan or its shorter analog Tp10), followed by hybridization to a protruding single-stranded portion of a double-stranded oligonucleotide decoy, and internalized in cultured cells [251]. Recently, there have been reports of the synthesis of peptide nucleic acids conjugated to either tri- and hepta-arginine or short basic Tat (48-57) peptides for the purpose of efficient cellular delivery of PNA [252]. The antisense effects of these kind of conjugated peptides are effectively enhanced by the use of calcium ions and lysosomotrophic agent Chloroquine. In fact, Nielsen and coworkers reported that the antisense activities of Tat(48-60) and heptaarginine-conjugated PNAs increased 44-fold and 8.5-fold, respectively, in the presence of 6mM Ca^{2+} ions suggesting an endosomal pathway for the internalization [166]. Moreover, octa(l-lysine) peptide covalently linked to PNA could facilitate free uptake into BCL1 cells as well as primary murine macrophages [167]. Taken together, all these approaches will probably be able to overcome the limitations of poor cellular delivery of PNA [253].

Another interesting method for *in vitro* cellular delivery of peptide nucleic acids is use of cationic lipid [238]. In the case of nucleic acids, the cationic lipid associates itself with the negatively charged phosphodiester backbone, fuses to the cell membrane and thereby allows the oligonucleotide to enter into the cell through an endocytotic pathway. According to an improvised version by Corey and coworkers [238,254], PNA oligomers are hybridized to overlapping oligonucleotides and the resulting complex is mixed with cationic lipid. The PNA-DNA-lipid complex is internalized and the partially hybridized PNA, which is imported into the cell as a passive cargo, expectedly dissociates itself from the complex [238,254]. A lipid-mediated delivery of PNAs pulmonary endothelium in mice [255] and the use of lipophilic molecules for an enhanced cellular uptake of PNA have been reported [173, 256,257]. The incorporation of PNA into delivery vehicles (or vesicles), like liposomes, has been widely adopted by many research groups. Liposomes have been considered as an *in vitro* model for cell membranes and the cell penetration of PNA molecules covalently attached to lipophilic adamantyl groups has been investigated [258,258b]. It was found out that the adamantyl modification increases the membrane-penetration rate of PNA by three-fold, as compared to corresponding unmodified PNA [258]. PNAs conjugated to adamantyl acetic acid have been used to study the liposome-mediated uptake in human cell lines like HeLa, IMR-90 and MDA-MB-453. The presence of adamantly moiety greatly improved (endosomal) cellular uptake, the rate of which is also dependent on the cell line. Moreover, the uptake of such lipophilic PNAs is further mediated by cationic liposomes [259]. Adamantly-conjugated PNA molecules along with cationic liposomes were used to achieve efficient delivery (~5-fold with respect to control) of PNAs into promyelocytic leukemia NB4 cells [260]. The cationic liposomes are also efficient vehicles for DNA-PNA hybrids [261,262]. Liposomes can also serve as protective agents for PNA-DNA chimera in severe experimental conditions that would otherwise comprise its stability. In principle, since double-stranded PNA-DNA-PNA chimeras bind to cationic liposomes, their stability in the presence of high concentrations of serum and cellular extracts actually increases as a result of such binding [263]. Novel methods for covalently attaching PNA molecules to liposomes are being investigated [121].

Conclusions

Peptide nucleic acids constitute a very interesting class of compounds with great potentials. The chemical and biological stability of PNAs and their sequence-specific binding to nucleic acids make them promising therapeutic lead compounds as well as novel agents for diagnostic applications. Over the years, developments have been made in many areas of chemistry, biology and biotechnology with PNA as a recognition molecule, to be used as a novel tool for diagnostic applications [9,77,154,208,209]. Moreover, based on PNA selectivity, recently a theoretical statistical mechanical model for gene targeting has been developed, which could provide a basis for strategic considerations in gene therapeutic context [264]. However, further development of more efficient and safe methods for its uptake and cell penetration is required before PNA could be used as a generic therapeutic drug. It appears that conjugated PNA molecules are better candidates for this purpose. Several efforts have been made to improve the applications of PNA, including further modifications of backbone and the development of second and third generations of PNA molecules and novel base analogues [36, 265-279].

References

[1] Nielsen, P. E., Egholm, M., Berg, R. H. and Buchardt, O. (1991) Sequence selective recognition of DNA by strand displacement with a thymine substituted polyamide. *Science* 254, 1497-1500.

[2] Egholm, M., Buchardt, O., Nielsen, P. E. and Berg, R. H. (1992) Peptide nucleic acids (PNA). Oligonucleotide analogues with an achiral peptide backbone. *J. Am. Chem. Soc.* 114, 1895-1897.

[3] Egholm, M., Buchardt, O., Christensen, L., Behrens, C., Frier, S. M., Driver, D. A., Berg, R. H., Kim, S. K, Nordén, B. and Nielsen, P. E. PNA hybridizes to complementary oligonucleotides obeying the Watson-Crick hydrogen bonding rules. *Nature* 1993; 365:566-568.

[4] Egholm, M., Nielsen, P. E., Buchardt, O. and Berg, R. H. (1992) Recognition of guanine and adenine in DNA by thymine and cytosine containing peptide nucleic acids. *J. Am. Chem. Soc.* 114, 9677-9678.

[5] Demidov, V., Frank-Kamenetskii, M. D., Egholm, M., Buchardt, O. and Nielsen, P. E. (1993) Sequence specific double strand DNA cleavage by Peptide Nucleic Acid (PNA) targeting using nuclease S1. *Nucleic Acids Res.* 21, 2103-2107.

[6] Demers, D. B., Curry, E. T., Egholm, M and Sozer, A. C. (1995) Enhanced PCR amplification of VNTR locus D1S80 using peptide nucleic acid (PNA). *Nucleic Acids Res.*, 23, 3050-3055.

[7] Perry-O'Keefe, H., Yao, X-W., Coull, J. M., Fuchs, M. and Egholm, M. (1996) Peptide nucleic acid pre-gel hybridization: An alternative to Southern hybridization. *Proc. Natl. Acad. Sci. U. S. A.* 93, 14670-14675.

[8] Veselkov, A. G., Demidov, V., Nielsen, P. E. and Frank-Kamenetskii, M. D. (1996) A new class of genome rare cutters. *Nucleic Acids Res.* 24, 2483-2487.

[9] Ray, A; Norden, B. Peptide nucleic acid (PNA): its medical and biotechnical applications and promise for the future: *FASEB J* 2000;14:1041-1060.

[10] Hanvey, J. C., Peffer, N. C., Bisi, J. E., Thomson, S. A., Cadilla, R., Josey, J. A., Ricca, D. J., Hassman, C. F., Bonham, M. A., Au, K. G., Carter, S. G., Bruckenstein, D. A., Boyd, A. L., Noble, S. A. and Babiss, L. E. (1992) Antisense and antigene properties of peptide nucleic acids. *Science* 258, 1481-1485.

[11] Nielsen, P. E., Egholm, M., Berg, R. H. and Buchardt, O. (1993) Peptide nucleic acids (PNA). Poteintial antisense and antigene agents. *Anti Cancer Drug Design* 8, 53-63.

[12] Boffa, L. C., Morris, P. L., Carpeneto, E. M., Louissaint, M. and Allfrey, V. G. (1996) Invasion of the CAG triplet repeats by a complementary peptide nucleic acid inhibits transcription of the androgen receptor and TATA-binding protein genes and correlates with refolding of an active nucleosome containing a unique AR gene sequence. *J. Biol. Chem.* 271, 13228-13233.

[13] Örum, H., Nielsen, P. E., Egholm, M., Berg, R. H., Buchardt, O. and Stanley, C. (1993) Single base pair mutation analysis by PNA directed PCR clamping. *Nucleic Acids Res.* 21, 5332-5336.

[14] Carlsson, C., Jonsson., M., Nordén, B., Dulay, M. T., Zare, R. N., Noolandi, J., Nielsen, P. E., Tsui, L-C. and Zielenski, J. Screening for genetic mutations. *Nature* 1996; 380:, 207.

[15] Thiede, C., Bayerdörffer, E., Blasczyk, R, Wittig, B. and Neubauer, A. (1996) Simple and sensitive detection of mutations in the ras proto-oncogenes using PNA-mediated PCR clamping. *Nucleic Acids Res.* 24, 983-984.

[16] Wang, J., Palecek, E., Nielsen, P. E., Rivas, G., Cai, X., Shirashi, H., Dontha, N., Luo, D. and Farias, P. A. M. (1996) Peptide Nucleic Acid Probes for Sequence-Specific DNA Biosensors. *J. Am. Chem. Soc.*118, 7667-7670.

[17] Nielsen, P. E. (1998) Structural and biological properties of peptide nucleic acid (PNA). *Pure and Appl. Chem.* 70, 105-110.

[18] Nielsen, P. E. (1998) Sequence-specific recognition of double-stranded DNA by peptide nucleic acids. Advances in DNA Sequence-Specific Agents, vol. 3, 267-278.

[19] Corey, D. R. (1997) Peptide nucleic acids: expanding the scope of nucleic acid recognition. *Trends Biotechnol.* 15, 224-229.

[20] Eriksson, M. and Nielsen, P. E. (1996) PNA-nucleic acid complexes: Structure, stability and dynamics. *Quarterly Rev. Biophysics* 29, 369-394.

[21] Merrifield, R. B. (1963) Solid-phase synthesis. I. The synthesis of a tetrapeptide. *J. Am. Chem. Soc.* 85, 2149-2154.

[22] Merrifield, B. (1986) Solid-phase synthesis. *Science* 232, 341-347.

[23] Christensen, L., Fitzpatrick, R., Gildea, B., Petersen, K. H., Hansen, H. F., Koch, T., Egholm, M., Buchardt, O., Nielsen, P. E., Coull, J. and Berg, R. H. (1995) Solid-phase Synthesis of Peptide Nucleic Acids. *J. Peptide Sc.* 3, 175-183.

[24] Norton, J. C., Waggenspack, J. H., Varnum, E. and Corey, D. R. (1995) Targeting peptide nucleic acid-protein conjugates to structural features within duplex DNA. *Bioorg. Med. Chem.* 3, 437-445.

[25] Haaima, G., Lohse, A., Buchardt, O. and Nielsen, P. E. (1996) Peptide Nucleic Acids (PNAs) Containing Thymine Monomers Derived from Chiral Amino Acids: Hybridization and Solubility Properties of D-lysine PNA. *Angew. Chem. Int. Ed. Engl.* 35, 1939-1942.

[26] Dueholm, K., Egholm, M., Behrens, C., Christensen, L., Hansen, H. F., Vulpius, T., Petersen, K. H., Berg, R. H., Nielsen, P. E. and Buchardt, O. (1994) Synthesis of Peptide Nucleic-Acid Monomers Containing the 4 Natural Nucleobases - Thymine,

Cytosine, Adenine, and Guanine and their Oligomerization. *J. Org. Chem.* 59, 5767-5773.

[27] Thomson, S. A., Josey, J. A., Cadilla, R., Gaul, M. D., Hassman, C. F., Luzzio, M. J., Pipe, A. J., Reed, K. L., Ricca, D. J., Wiethe, R. W. and Noble, S. A. (1995) Fmoc Mediated Synthesis of Peptide Nucleic-Acids. *Tetrahedron* 51, 6179-6194.

[28] Egholm, M., Christensen, L., Dueholm, K. L., Buchardt, O., Coull, J. and Nielsen, P. E. (1995) Efficient pH-independent sequence-specific DNA-binding by pseudoisocytosine-containing bis-PNA. *Nucleic Acids Res.* 23, 217-222.

[29] Uhlmann E., Peyman, A., Breipohl, G. and Will D. W. (1998) PNA: Synthetic polyamide nucleic acids with unusual binding properties. *Angew. Chem. Int. Ed.* 37, 2796-2823.

[30] De Koning, MC; Van der Marel, GA; and Overhand M. Synthetic developments towards PNA–peptide conjugates. *Current Opinion in Chemical Biology* 2003, 7:734–740.

[31] Koch, T., Naesby, M., Wittung, P., Joergensen, M., Larsson, C., Buchardt, O., Stanley, C. J., Nordén, B. Nielsen, P. E. and Örum, H. (1995) PNA-peptide chimerae. *Tetrahedron Lett.* 36, 6933-6936.

[32] Betts, L., Josey, J. A., Veal, J. M. and Jordan, S. R. (1995) A Nucleic Acid Triple Helix formed by a Peptide Nucleic Acid-DNA Complex. *Science*, 270, 1838-1841.

[33] Awasthi SK, Nielsen PE: Parallel synthesis of PNA-peptide conjugate libraries. *Comb Chem High Throughput Screen* 2002, 5:253-259.

[34] Kaihatsu K, Corey DR; Synthesis of peptide nucleic acid-peptide conjugates, .*Methods Mol Biol.* 2004;283:207-14.

[35] Kaushik N, Basu A, Palumbo P, Myers RL, Pandey VN: Anti-TAR polyamide nucleotide analog conjugated with a membrane permeating peptide inhibits human immunodeficiency virus type 1 production. *J Virol* 2002, 76:3881-3891.

[36] Amirkhanov, NV; Wickstrom, E; Synthesis of novel polydiamidopropanoate dendrimer PNA-peptide chimeras for non-invasive magnetic resonance imaging of cancer. *Nucleosides Nucleotides Nucleic Acids*. 2005;24, :423-426.

[37] Butler, J. M., Jing-Baucom, P., Huang, M., Belgrader, P., and Gairard, J. (1996) Peptide Nucleic Acid characterization by MALDI-TOF Mass Spectrometry. *Anal. Chem.*, 68, 3283-3287.

[38] Ross, P. L., Lee, K. and Belgrader, P. (1997) Discrimination of single-nucleotide polymorphisms in human DNA using peptide nucleic acid probes detected by MALDI-TOF mass spectrometry. *Anal. Chem.*, 69, 4197-4202.

[39] Tomac, S., Sarkar, M., Ratilainen, T., Wittung, P., Nielsen, P. E., Nordén, B. and Gräslund, A. Ionic effects on the stability and conformation of Peptide Nucleic acid complexes. *J Am. Chem. Soc.* 1996; 118:5544-5552.

[40] Kuhn, H., Demidov, V. V., Frank-Kamenetskii, M. D. and Nielsen, P. E. (1998) Kinetic sequence discrimination of cationic bis-PNAs upon targeting of double-stranded DNA. *Nucleic Acids Res.* 26, 582- 587.

[41] Eriksson, M., Christensen, L., Schmidt, J., Haaima, G., Orgel, L and Nielsen, P. E. (1998) Sequence dependent N-terminal rearrangement and degradation of peptide nucleic acid (PNA) in aqueous solution. *New J. Chem.* 1055-1059.

[42] Hyrup, B. and Nielsen, P. E. (1996) Peptide nucleic acids (PNA): synthesis, properties and potential applications. *Bioorg. Med. Chem.* 4, 5-23.

[43] Riguet E, Tripathi S, Chaubey B, Desire J, Pandey VN, Decout JL. A peptide nucleic acid-neamine conjugate that targets and cleaves HIV-1 TAR RNA inhibits viral replication. *J Med Chem.* 2004,47, 4806-4809.

[44] Nielsen, P. E., Egholm, M. and Buchardt, O. (1994) Evidence for (PNA)2/DNA triplex structure upon binding of PNA to dsDNA by strand displacement. *J. Mol. Recognition* 7, 164-170.

[45] Cherney, D. Y., Belotserkovskii, B. P., Frank-Kamenetskii, M. D., Egholm, M., Buchardt, O., Berg, R. H. and Nielsen, P. E. (1993) DNA unwinding upon strand-displacement binding of a thymine-substituted polyamide to double-stranded DNA. *Proc. Natl. Acad. Sci. U. S. A.* 90, 1667-1670.

[46] Peffer, N. J., Hanvey, J. C., Bisi, J. E., Thomson, S. A., Hassman, C. F., Noble, S. A. and Babiss, L. E. (1993) Strand-invasion of duplex DNA by peptide nucleic acid oligomers. *Proc. Natl. Acad. Sci. U. S. A.* 90, 10648-10652.

[47] Wittung, P., Nielsen, P. E and Nordén, B. Direct observation of strand invasion by peptide nucleic acid (PNA) into double-stranded DNA. *J. Am. Chem. Soc.* 1996;118:7049-7054. 47b. Wittung, P., P. Nielsen and Nordén, B. Recognition of Double-Stranded DNA by Peptide Nucleic Acids *Nucleosides and Nucleotides* 1997; 16:599-602. 47c. Hyrup, B., Egholm, M., Nielsen, P. E., Wittung, P., Nordén, B. and Buchardt, O. Structure-Activity Studies of the Binding of Modified Peptide Nucleic Acids (PNAs) to DNA . *J. Am. Chem. Soc.* 1994;116:7964-7970.

[48] Jensen, K. K., Örum, H., Nielsen, P. E. and Nordén, B. Kinetics for hybridization of peptide nucleic acids (PNA) with DNA and RNA studied with the BIAcore technique. *Biochemistry* 1997; 36:5072-5077.

[49] Esposito, V; Galeone, A; Mayol, L; Messere, A, Piccialli, G; Randazzo, A. PNA-DNA chimeras forming quadruplex structures. *Nucleosides Nucleotides Nucleic Acids.* 2003; 22:1681-1684.

[50] Bajor, Z; Sagi, G; Tegyey, Z; Otvos, L. Synthesis, biophysical, and biochemical properties of PNA-DNA chimeras. *Nucleosides Nucleotides Nucleic Acids.* 2003; 22:1215-1217.

[51] Wittung, P., Nielsen, P. E., Buchardt,. O., Egholm, M. and Nordén, B. DNA like Double Helix formed by Peptide Nucleic Acid. *Nature* 1994; 368:561-563.

[52] Wittung, P., Eriksson, M., Lyng, R., Nielsen, P. E. and Nordén, B. Induced Chirality in PNA-PNA Duplexes. *J. Am. Chem. Soc.* 1995; 117:10167-10173.

[53] Wittung, P., Nielsen, P. and Nordén, B. Extended DNA-recognition repertoire of PNA. *Biochemistry* 1997; 36:7973-7979. 53b. Wittung, P., Nielsen, P. and Nordén, B. Observation of PNA-PNA-PNA triplex *J. Amer. Chem. Soc* 1997;119:3189-3190.

[54] Leijon, M., Gräslund, A., Nielsen, P. E., Buchardt, O., Nordén, B., Kristensen, S. M. and Eriksson, M. (1994) Structural Characterization of PNA-DNA duplexes by NMR. Evidence for DNA in a B-like Conformation. *Biochemistry*, 33, 9820-9825.

[55] Eriksson, M. and Nielsen, P. E. (1996) Solution Structure of a Peptide Nucleic Acid-DNA duplex. *Nature Struct. Biol.* 3, 410-413.

[56] Brown, S. C., Thomson, S. A., Veal, J. M. and Davis, D. G. (1994) NMR Solution Structure of a Peptide Nucleic Acid Complexed with RNA. *Science* 265, 777-780.

[57] Rasmussen, H., Kastrup, J. S., Nielsen, J. N., Nielsen, J. M. and Nielsen, P. E. (1997) Crystal Structure of a Peptide Nucleic Acid (PNA) Duplex at 1.7 Å Resolution. *Nature Struct. Biol.* 4, 98-101.

[58] Wittung, P; Kim, SK; Buchardt, O; Nielsen, PE; and Nordén B. Interactions of DNA binding ligands with PNA-DNA hybrids. *Nucleic Acids Res.*, 1994; 22:5371-5377.

[59] Ratilainen, T., Holmén, A., Tuite, E., Haaima, G., Christensen, L., Nielsen, P. E. and Nordén, B. Hybridization of Peptide Nucleic Acid. *Biochemistry* 1998:37: 12331-12342.

[60] Holmén. A. and Nordén, B. (1999) Thermodynamics of PNA-Nucleic Acid Interactions. In *Peptide Nucleic Acids: Protocols and Applications* (Nielsen, . P. E. and Egholm, M., eds) pp 87-97, Horizon Press, Wymondham, UK.

[61] Giesen, U., Kleider, W., Berding, C., Ørum, H. and Nielsen, P. E. (1998) A formula for thermal stability (T_m) prediction of PNA/DNA duplexes. *Nucleic Acids Res.* 26, 1601-1620.

[62] Ratilainen T; Norden B. Thermodynamics of PNA interactions with DNA and RNA. *Methods Mol Biol: Peptide Nucleic Acids (Ed. P.E. Nielsen). Humana Press, Totowa, New Jersey,* 2002;208:59-88.

[63] Krupnik OV, Guscho Y, Sluchanko K, Nielsen P, Lazurkin Y. Thermodynamics of the melting of PNA(2)/DNA triple helices. *J Biomol Struct Dyn.* 2001: 19:535-542.

[64] Kushon SA, Jordan JP, Seifert JL, Nielsen H, Nielsen PE, Armitage BA. Effect of secondary structure on the thermodynamics and kinetics of PNA hybridization to DNA hairpins. *J Am Chem Soc.* 2001: 123:10805-10813.

[65] Sugimoto N, Satoh N, Yasuda K, Nakano S. Stabilization factors affecting duplex formation of peptide nucleic acid with DNA. *Biochemistry.* 2001: 40:8444-8451.

[66] Ratilainen T, Holmen A, Tuite E, Nielsen PE, Norden B. Thermodynamics of sequence-specific binding of PNA to DNA. *Biochemistry.* 2000: 39:7781-7791.

[67] Chakrabarti MC; Schwarz FP. Thermal stability of PNA/DNA and DNA/DNA duplexes by differential scanning calorimetry. *Nucleic Acids Res.* 1999: 27:4801-4806.

[68] Schwarz FP; Robinson S; Butler JM. Thermodynamic comparison of PNA/DNA and DNA/DNA hybridization reactions at ambient temperature. *Nucleic Acids Res.* 1999: 27:4792-4800.

[69] Sen A; Nielsen PE. Unique properties of Purine/Pyrimidine asymmetric PNADNA duplexes: Differential stabilization of PNADNA duplexes by purines in the PNA strand. *Biophys J.* 2005: Dec 2; [Epub ahead of print].

[70] Kim, S. K., Nielsen, P. E., Egholm, M., Buchardt, O., Berg, R. H. and Nordén, B. Right-Handed Triplex formed between Peptide Nucleic Acid PNA-T_8 AND poly(dA) shown by Linear and Circular Dichroism Spectroscopy. *J. Am. Chem. Soc.* 1993; 115: 6477-6481.

[71] Knudsen, H. and Nielsen, P. E. (1996) Antisense properties of duplex- and triplex-forming PNAs. *Nucleic Acids Res.* 24, 494-500.

[72] Nielsen, P. E. (1999) Peptide nucleic acids as therapeutic agents. *Curr. Opin. Struct. Biol.* 9, 353-357.

[73] Abibi A, Protozanova E, Demidov VV, Frank-Kamenetskii MD,. Specific versus nonspecific binding of cationic PNAs to duplex DNA. *Biophys J.* 2004; 86:3070-3078.

[74] Sen, S. and Nilsson, L. (1998) Molecular Dynamics of Duplex Systems Involving PNA: Structural and Dynamical Consequences of the Nucleic Acid Backbone. *J. Am. Chem. Soc.* 120, 619-631

[75] Shields, G. C., Laughton, C. A. and Orozco, M. (1998) Molecular Dynamics Simulation of a PNA.DNA.PNA Triple Helix in Aqueous Solution. *J. Am. Chem. Soc.* 120, 5895-5904.

[76] Srinivasan, A. R. and Olson, W. K. (1998) Molecular Models of Nucleic Acid Triple Helixes. II. PNA and 2′-5′ Backbone Complexes. *J. Am. Chem. Soc.* 120, 492-499

[77] Petersson B, Nielsen BB, Rasmussen H, Larsen IK, Gajhede M, Nielsen PE, Kastrup, JS. . Crystal structure of a partly self-complementary peptide nucleic acid (PNA) oligomer showing a duplex-triplex network. *J Am Chem Soc.* 2005; Feb. 127:1424-1430.

[78] Rathinavelan, T; Yathindra N. Molecular dynamics structures of peptide nucleic acid x DNA hybrid in the wild-type and mutated alleles of Ki-ras proto-oncogene--stereochemical rationale for the low affinity of PNA in the presence of an AC mismatch.*FEBS J.* 2005; 272:4055-4070.

[79] Olsen, AG; Dahl, O; Nielsen, PE. Synthesis and evaluation of a conformationally constrained pyridazinone PNA-monomer for recognition of thymine in triple-helix structures. *Bioorg Med Chem Lett.* 2004; 14:1551-1554.

[80] Petraccone L, Erra E, Messere A, Montesarchio D, Piccialli G, De Napoli L, Barone G, Giancola C. Targeting duplex DNA with DNA-PNA chimeras? Physico-chemical characterization of a triplex DNA-PNA/DNA/DNA. *Biopolymers.* 2004;73:434-442.

[81] Rasmussen, H; Liljefors, T; Petersson, B; Nielsen, PE; Liljefors, T; Kastrup, JS. The influence of a chiral amino acid on the helical handedness of PNA in solution and in crystals. *J Biomol Struct Dyn.* 2004; 21:495-502.

[82] Menchise, V; De Simone, G; Tedeschi, T; Corradini, R; Sforza, S; Marchelli, R; Capasso, D; Saviano, M; Pedone, C. Insights into peptide nucleic acid (PNA) structural features: the crystal structure of a D-lysine-based chiral PNA-DNA duplex. *Proc Natl Acad Sci U S A.* 2003; 100:12021-12026.

[83] Govindaraju, T; Gonnade, RG; Bhadbhade, MM; Kumar, VA; Ganesh KN. (1S,2R/1R,2S)-ainocyclohexyl glycyl thymine PNA: synthesis, monomer crystal structures, and DNA/RNA hybridization studies. *Org Lett.* 2003; 5:3013-3036.

[84] Christensen, C; Eldrup, AB; Haaima, G; Nielsen, PE. 1,8-Naphthyridin-2,7-(1,8H)-dione is an effective mimic of protonated cytosine in peptide nucleic acid triplex recognition systems. *Bioorg Med Chem Lett.* 2002; 12:3121-3124.

[85] Krupnik, OV; Guscho, Y; Sluchanko, K; Nielsen, P; Lazurkin, Y. Thermodynamics of the melting of PNA(2)/DNA triple helices. *J Biomol Struct Dyn.* 2001; 19:535-542.

[86] Datta, B; Armitage, BA; Hybridization of PNA to structured DNA targets: quadruplex invasion and the overhang effect. *J Am Chem Soc.* 2001; 123:9612-9619.

[87] Datta, B; Schmitt, C; Armitage, BA. Formation of a PNA2-DNA2 hybrid quadruplex. *J Am Chem Soc.* 2003; 125: 4111-4118.

[88] Esposito, V; Galeone, A; Mayol, L; Messere, A; Piccialli, G; Randazzo, A. PNA-DNA chimeras forming quadruplex structures. *Nucleosides Nucleotides Nucleic Acids*. 2003; 22: 1681-1684.

[89] Datta, B; Bier, ME; Roy, S; Armitage, BA. Quadruplex formation by a guanine-rich PNA oligomer. *J Am Chem Soc*. 2005 127:4199-4207.

[90] Krishnan-Ghosh, Y; Whitney, AM; Balasubramanian S. Dynamic covalent chemistry on self-templating PNA oligomers: formation of a bimolecular PNA quadruplex. *Chem Commun (Camb)*. 2005; 24: 3068-3070.

[91] Petraccone, L; Pagano, B; Esposito, V; Randazzo, A; Piccialli, G; Barone, G; Mattia, CA; Giancola, C. Thermodynamics and kinetics of PNA-DNA quadruplex-forming chimeras. *J Am Chem Soc*. 2005; 127:16215-16223.

[92] Misra, H. S., Pandey, P. K., Modak, M. J., Vinayak, R. and Pandey, V. N. Polyamide Nucleic Acid-DNA chimera lacking the phosphate backbone are novel primers for polymerase reaction catalyzed by DNA polymerases. *Biochemistry* 1998; 37:1917-1925.

[93] Koppitz, M., Nielsen, P. E. and Orgel, L. E. Formation of oligonucleotide-PNA-chimeras by template-directed ligation. *J. Am Chem. Soc*. 1998; 120;4563-4569.

[94] Orum H. PCR clamping. *Curr Issues Mol Biol*. 2000; 2:27-30. 94a.Murdock DG, Wallace DC. PNA-mediated PCR clamping. Applications and methods. *Methods Mol Biol*. 2002;208:145-64.

[95] Rhodes CH, Honsinger C, Porter DM, Sorenson GD Analysis of the allele-specific PCR method for the detection of neoplastic disease.*Diagn Mol Pathol*. 1997; 6:49-57.

[96] Behn M, Thiede C, Neubauer A, Pankow W, Schuermann M Facilitated detection of oncogene mutations from exfoliated tissue material by a PNA-mediated 'enriched PCR' protocol. *J Pathol*. 2000; 190:69-75.

[97] Takahashi S, Chi XJ, Yamaguchi Y, Suzuki H, Sugaya S, Kita K, Hiroshima K, Yamamori H, Ichinose M, Suzuki N. Mutagenicity of bisphenol A and its suppression by interferon-alpha in human Rsa cells. *Mutat Res*. 2001;490:199-207.

[98] Behn M, Schuermann M. Simple and reliable factor V genotyping by PNA-mediated PCR clamping.*Thromb Haemost*. 1998; 79: 773-777.

[99] Cochet O, Martin E, Fridman WH, Teillaud JL. Selective PCR amplification of functional immunoglobulin light chain from hybridoma containing the aberrant MOPC 21-derived V kappa by PNA-mediated PCR clamping. *Biotechniques*. 1999; 26: 818-820, 822.

[100] Zhong S, Nguyen NY, Eggerman TL. Detection of apolipoprotein B mRNA editing by peptide nucleic acid mediated PCR clamping. *Biochem Biophys Res Commun*. 1999; 259:311-313.

[101] Murdock DG, Christacos NC, Wallace DC. The age-related accumulation of a mitochondrial DNA control region mutation in muscle, but not brain, detected by a sensitive PNA-directed PCR clamping based method. *Nucleic Acids Res*. 2000;28::4350-4355.

[102] Fiandaca MJ, Hyldig-Nielsen JJ, Gildea BD, Coull JM. Self-reporting PNA/DNA primers for PCR analysis. *Genome Res*. 2001; 11:609-613.

[103] Alsmadi OA, Bornarth CJ, Song W, Wisniewski M, Du J, Brockman JP, Faruqi AF, Hosono S, Sun Z, Du Y, Wu X, Egholm M, Abarzua P, Lasken RS, Driscoll MD. High accuracy genotyping directly from genomic DNA using a rolling circle amplification based assay. *BMC Genomics*. 2003; 30;4:21.

[104] Prix L, Uciechowski P, Bockmann B, Giesing M, Schuetz AJ. Diagnostic biochip array for fast and sensitive detection of K-ras mutations in stool. *Clin Chem*. 2002;48:428-435.

[105] Kreuzer KA, Le Coutre P, Landt O, Na IK, Schwarz M, Schultheis K, Hochhaus A, Dorken B. Preexistence and evolution of imatinib mesylate-resistant clones in chronic myelogenous leukemia detected by a PNA-based PCR clamping technique. *Ann Hematol*. 2003; 82:284-289.

[106] Horny HP, Lange K, Sotlar K, Valent P. Increase of bone marrow lymphocytes in systemic mastocytosis: reactive lymphocytosis or malignant lymphoma? Immunohistochemical and molecular findings on routinely processed bone marrow biopsy specimens. *J Clin Pathol*. 2003; 56:575-578.

[107] Dabritz J, Hanfler J, Preston R, Stieler J, Oettle H. Detection of Ki-ras mutations in tissue and plasma samples of patients with pancreatic cancer using PNA-mediated PCR clamping and hybridisation probes. *Br J Cancer*. 2005;92:405-412.

[108] Lietman SA, Ding C, Levine MA. A highly sensitive polymerase chain reaction method detects activating mutations of the GNAS gene in peripheral blood cells in McCune-Albright syndrome or isolated fibrous dysplasia *J Bone Joint Surg Am*. 2005; 87:2489-2494.

[109] Takiya T, Futo S, Tsuna M, Namimatsu T, Sakano T, Kawai K, Suzuki T. Identification of single base-pair mutation on uidA gene of Escherichia coli O157:H7 by Peptide Nucleic Acids (PNA) mediated PCR clamping. *Biosci Biotechnol Biochem*. 2004; 68:360-368.

[110] Takiya T, Horie Y, Futo S, Matsumoto Y, Kawai K, Suzuki T. Rapid selection of nonhotspot mutants among hisD+ revertants of Salmonella typhimurium TA98 in Ames test by peptide nucleic acid (PNA)-mediated PCR clamping. *J Biosci Bioeng*. 2003;96:588-590.

[111] Peano C, Lesignoli F, Gulli M, Corradini R, Samson MC, Marchelli R, Marmiroli N. Development of a peptide nucleic acid polymerase chain reaction clamping assay for semiquantitative evaluation of genetically modified organism content in food. *Anal Biochem*. 2005;344:174-182.

[112] Igloi, G. L. (1998) Variability in the stability of DNA-peptide nucleic acid (PNA) single-base mismatched duplexes: Real-time hybridization during affinity electrophoresis in PNA containing gels. *Proc. Natl. Acad Sci. U. S. A*. 95, 8562-8567

[113] Igloi GL Automated detection of point mutations by electrophoresis in peptide-nucleic acid-containing gels. *Biotechniques*. 1999;27:798-800, 802, 804 passim.

[114] Igloi GL.. Simultaneous identification of mutations by dual-parameter multiplex hybridization in peptide nucleic acid-containing virtual arrays. *Genomics*. 2001;74:402-407.

[115] Igloi GL. Detection of point mutations using PNA-containing electrophoresis matrices. *Methods Mol Biol*. 2002;208:195-207.

[116] Jansen K, Richelson E. Detection of peptide nucleic acids in tissue extracts of treated animals by gel mobility shift assay. *J Biochem Biophys Methods.* 2000; 42:31-34.

[117] Basile A, Giuliani A, Pirri G, Chiari M Use of peptide nucleic acid probes for detecting DNA single-base mutations by capillary electrophoresis *Electrophoresis.* 2002;23:926-929.

[118] Galaverna G, Sforza S, Tedeschi T, Corradini R, Dossena A, Marchelli R. Enantiomeric separation of chiral peptide nucleic acid monomers by capillary electrophoresis with charged cyclodextrins. *Electrophoresis.* 2003;24::2698-2703.

[119] Tedeschi T, Chiari M, Galaverna G, Sforza S, Cretich M, Corradini R, Marchelli R. Detection of the R553X DNA single point mutation related to cystic fibrosis by a "chiral box" D-lysine-peptide nucleic acid probe by capillary electrophoresis.*Electrophoresis.* 2005; 26:4310-4316.

[120] Rose, D. J. (1993) Characterization of antisense binding properties of peptide nucleic acids by capillary gel electrophoresis. *Anal. Chem.* 65, 3545-3549.

[121] BF, Schneider JW Sequence-specific binding of DNA to liposomes containing di-alkyl peptide nucleic acid (PNA) amphiphiles. *Langmuir.* 2005;21:2488-2494.

[122] Oehlke J, Wallukat G, Wolf Y, Ehrlich A, Wiesner B, Berger H, Bienert M. Enhancement of intracellular concentration and biological activity of PNA after conjugation with a cell-penetrating synthetic model peptide. *Eur J Biochem.* 2004; 271:3043-3049.

[123] BIAcore X Instrument Handbook(1996)., Preliminary Edition, Pharmacia Biosensor AB.

[124] Sawata S, Kai E, Ikebukuro K, Iida T, Honda T, Karube I. Application of peptide nucleic acid to the direct detection of deoxyribonucleic acid amplified by polymerase chain reaction *Biosens Bioelectron.* 1999;14:397-404.

[125] Fukuda N, Furuya R, Kishioka H, Suzuki R, Matsuda H, Tahira Y, Takagi H, Ikeda Y, Saito S, Matsumoto K, Kanmatsuse K J Effects of antisense peptide nucleic acid to platelet-derived growth factor A-chain on growth of vascular smooth muscle cells. *Cardiovasc Pharmacol.* 2003;42:224-231.

[126] Masuko M. Hybridization of an immobilized PNA probe with its complementary oligodeoxyribonucleotide on the surface of silica glass. *Nucleic Acids Res Suppl.* 2003;3:145-146.

[127] Corradini R, Feriotto G, Sforza S, Marchelli R, Gambari R.. Enhanced recognition of cystic fibrosis W1282X DNA point mutation by chiral peptide nucleic acid probes by a surface plasmon resonance biosensor. *J Mol Recognit.* 2004;17:76-84.

[128] Endo T, Kerman K, Nagatani N, Takamura Y, Tamiya E. Label-free detection of peptide nucleic acid-DNA hybridization using localized surface plasmon resonance based optical biosensor. *Anal Chem.* 2005;77:6976-6984.

[129] Brandt O, Hoheisel JD. Peptide nucleic acids on microarrays and other biosensors. *Trends Biotechnol.* 2004;22::617-622.

[130] Wang, J., Nielsen, P. E., Jiang, M., Cai, X., Fernandes, J. R., Grant, D. H., Ozsoz, M., Beglieter, A. and Mowat, M. (1997) Mismatch-sensitive hybridization detection by peptide nucleic acids immobilized on a Quartz-Crystal Microbalance. *Anal. Chem.* 69, 5200-5202

[131] Höök, F; Ray, A; Nordén, B; Kasemo, B Characterization of PNA and DNA Immobilization and Subsequent Hybridization with DNA Using Acoustic-Shear-Wave Attenuation Measurements. *Langmuir* 2001;17:8305 –8312.

[132] Kirpekar, F., Nordhoff, E., Larsen, K. K., Kristiansen, K, Roepstorff, P. and Hillenkamp, F. (1998) DNA Sequence Analysis by MALDI Mass Spectrometry. *Nucleic Acids Res.* 26, 2554-2559.

[133] Berkenkamp, S., Kirpekar, F. and Hillenkamp, F. (1998) Infrared MALDI mass spectrometry of large nucleic acids. *Science* 281, 260-262.

[134] Egholm, M. (1997) Spectrometry Senses more than a Small Difference. *Nature Biotechnol.* 15, 1346.

[135] Griffin, T. J., Tang, W. and Smith, L. M. (1997) Genetic Analysis by Peptide Nucleic Acid Affinity MALDI-TOF Mass Spectrometry. *Nature Biotechnol.* 15, 1368-1372.

[136] Brandt O, Hoheisel JD Peptide nucleic acids on microarrays and other biosensors. *Trends Biotechnol.* 2004; 22:617-622.

[137] Wilhelmsson LM, Norden B, Mukherjee K, Dulay MT, Zare RN. Genetic screening using the colour change of a PNA-DNA hybrid-binding cyanine dye. *Nucleic Acids Res.* 2002;30:E3.

[138] Nielsen, P. E., Egholm, M. and Buchardt, O. (1994) Sequence-specific transcription arrest by peptide nucleic acid bound to the DNA template strand. *Gene* 149, 139-145.

[139] Praseuth, D., Grigoriev, M., Guieysse, A. L., Pritchard, L. L., Harel-Bellan, A., Nielsen, P. E. and Helene, C. (1996) Peptide nucleic acids directed to the promoter of the alpha-chain of the interleukin-2 receptor. *Biochim. Biophys. Acta* 1309, 226-238.

[140] Bentin, T. and Nielsen, P. E. (1996) Enhanced Peptide Nucleic acid binding to supercoiled DNA: possible implications for DNA "breathing" dynamics".*Biochemistry* 35, 8863-8869.

[141] Kuhn, H., Demidov, V. V., Frank-Kamenetskii, M. D. and Nielsen, P. E. (1998) Kinetic sequence discrimination of cationic bis-PNAs upon targeting of double-stranded DNA. *Nucleic Acids Res.* 26, 582- 587.

[142] Armitage, B., Koch, T., Frydenlund, H., Örum, H. and Schuster, G. B. (1998) Peptide nucleic acid (PNA)/DNA hybrid duplexes: intercalation by an internally linked anthraquinone. *Nucleic Acids Res.* 26, 715-720.

[143] Larsen, H. J. and Nielsen, P. E. (1996) Transcription-mediated binding of peptide nucleic acid (PNA) to double-stranded DNA: sequence-specific suicide transcription. *Nucleic Acids Res.* 24, 458-463.

[144] Mollegaard, N. E., Buchardt, O., Egholm, M. and Nielsen, P. E. (1994) Peptide nucleic acid. DNA strand displacement loops as artificial transcription promoters. *Proc. Natl. Acad. Sci. U. S. A.* 91, 3892-3895.

[145] Lee, R., Kaushik, N., Modak, M. J., Vinayak, R. and Pandey, V. N. (1998) Polyamide nucleic acid targeted to the primer binding site of the HIV-1 RNA genome blocks *in vitro* HIV-1 reverse transcription. *Biochemistry* 37, 900-910

[146] McMahon BM, Stewart JA, Bitner MD, Fauq A, McCormick DJ, Richelson E. Peptide nucleic acids specifically cause antigene effects in vivo by systemic injection. *Life Sci.* 2002;71:325-337.

[147] McMahon BM, Stewart J, Fauq A, Younkin S, Younkin L, Richelson E. Using peptide nucleic acids as gene-expression modifiers to reduce beta-amyloid levels. *J Mol Neurosci.* 2002;19:71-76.

[148] Cogoi S, Rapozzi V, Xodo LE. Inhibition of gene expression by peptide nucleic acids in cultured cells. *Nucleosides Nucleotides Nucleic Acids.* 2003;22:1615-1618.

[149] Gambari R. Biological activity and delivery of peptide nucleic acids (PNA)-DNA chimeras for transcription factor decoy (TFD) pharmacotherapy. *Curr Med Chem.* 2004;11:1253-1263.

[150] Ziemba AJ, Zhilina ZV, Krotova-Khan Y, Stankova L, Ebbinghaus SW. Targeting and regulation of the HER-2/neu oncogene promoter with bis-peptide nucleic acids. Oligonucleotides. 2005;15:36-50.

[151] Kaihatsu K, Janowski BA, Corey DR. Recognition of chromosomal DNA by PNAs *Chem Biol.* 2004;11:749-758.

[152] Penolazzi L, Borgatti M, Lambertini E, Mischiati C, Finotti A, Romanelli A, Saviano M, Pedone C, Piva R, Gambari R. Peptide nucleic acid-DNA decoy chimeras targeting NF-kappaB transcription factors: Induction of apoptosis in human primary osteoclasts *Int J Mol Med.* 2004;14:145-152.

[153] Kaihatsu K, Shah RH, Zhao X, Corey DR. Extending recognition by peptide nucleic acids (PNAs): binding to duplex DNA and inhibition of transcription by tail-clamp PNA-peptide conjugates. *Biochemistry.* 2003;42;13996-14003.

[154] Nielsen PE. Gene targeting using peptide nucleic acid. *Methods Mol Biol.* 2005;288:343-358.

[155] Cogoi S, Codognotto A, Rapozzi V, Xodo LE. Antigene property of PNA conjugated to the nuclear localization signal peptide. *Nucleosides Nucleotides Nucleic Acids.* 2005;24:971-974.

[156] Mologni, L., leCoutre, P., Nielsen, P. E. and Gambacorti-Passerini, C. (1998) Additive antisense effects of different PNAs on the *in vitro* translation of the PML/RARα gene. *Nucleic Acids Res.* 26, 1934-1938.

[157] Good, L. and Nielsen, P. E. (1998) Inhibition of translation and bacterial growth by peptide nucleic acid targeted to ribosomal RNA. *Proc. Natl. Acad. Sci. U. S. A.* 95, 2073-2076.

[158] Good, L. and Nielsen, P. E. (1998) Antisense inhibition of gene expression in bacteria by PNA targeted to mRNA. *Nature Biotechnol.* 16, 355-358.

[159] Good L, Awasthi SK, Dryselius R, Larsson O, Nielsen PE Bactericidal antisense effects of peptide-PNA conjugates *Nat Biotechnol.* 2001;19:360-364.

[160] Nekhotiaeva N, Awasthi SK, Nielsen PE, Good L. Inhibition of Staphylococcus aureus gene expression and growth using antisense peptide nucleic acids *Mol Ther.* 2004;10:652-659.

[161] Kulyte A, Nekhotiaeva N, Awasthi SK, Good L. Inhibition of Mycobacterium smegmatis Gene Expression and Growth Using Antisense Peptide Nucleic Acids. *J Mol Microbiol Biotechnol.* 2005; 9:101-109.

[162] Shiraishi T, Nielsen PE Down-regulation of MDM2 and activation of p53 in human cancer cells by antisense 9-aminoacridine-PNA (peptide nucleic acid) conjugates. *Nucleic Acids Res.* 2004;32:4893-4902.

[163] Wickstrom E, Urtishak KA, Choob M, Tian X, Sternheim N, Cross LM, Rubinstein A, Farber SA. Downregulation of gene expression with negatively charged peptide nucleic acids (PNAs) in zebrafish embryos. *Methods Cell Biol.* 2004;77:137-158.

[164] Kilk K, Elmquist A, Saar K, Pooga M, Land T, Bartfai T, Soomets U, Langel U. Targeting of antisense PNA oligomers to human galanin receptor type 1 mRNA. *Neuropeptides*. 2004;38:316-324.

[165] Kaihatsu K, Huffman KE, Corey DR. Intracellular uptake and inhibition of gene expression by PNAs and PNA-peptide conjugates. *Biochemistry*. 2004;43:14340-14347.

[166] Shiraishi T, Pankratova S, Nielsen PE Calcium ions effectively enhance the effect of antisense peptide nucleic acids conjugated to cationic tat and oligoarginine peptides *Chem Biol.* 2005;12:923-929.

[167] Albertshofer K, Siwkowski AM, Wancewicz EV, Esau CC, Watanabe T, Nishihara KC, Kinberger GA, Malik L, Eldrup AB, Manoharan M, Geary RS, Monia BP, Swayze EE, Griffey RH, Bennett CF, Maier MA. Structure-activity relationship study on a simple cationic peptide motif for cellular delivery of antisense peptide nucleic acid. *J Med Chem.* 2005;48:6741-6749.

[168] Chiarantini L, Cerasi A, Fraternale A, Millo E, Benatti U, Sparnacci K, Laus M, Ballestri M, Tondelli L. Comparison of novel delivery systems for antisense peptide nucleic acids. *J Control Release*. 2005;109:24-36.

[169] Wilusz JE, Devanney SC, Caputi M. Chimeric peptide nucleic acid compounds modulate splicing of the bcl-x gene in vitro and in vivo. *Nucleic Acids Res.* 2005;33:6547-6554.

[170] Suzuki T, Zhang Y, Zhang YF, Schlachetzki F, Pardridge WM. Imaging gene expression in regional brain ischemia in vivo with a targeted [^{111}In]-antisense radiopharmaceutical.*Mol Imaging*. 2004;3:356-363.

[171] Taylor, R. W., Chinnery, P. F., Turnbull, D. M. and Lightowlers, R. N. (1997) Selective inhibition of mutant human mitochondrial DNA replication *in vitro* by peptide nucleic acids. *Nature Geneti.* 15, 212-215.

[172] Taylor RW, Chinnery PF, Turnbull DM, Lightowlers RN. In-vitro genetic modification of mitochondrial function. *Hum Reprod.* 2000;15 Suppl 2:79-85.

[173] Muratovska A, Lightowlers RN, Taylor RW, Turnbull DM, Smith RA, Wilce JA, Martin SW, Murphy MP. Targeting peptide nucleic acid (PNA) oligomers to mitochondria within cells by conjugation to lipophilic cations: implications for mitochondrial DNA replication, expression and disease. *Nucleic Acids Res.* 2001;29:1852-1863.

[174] Ross GF, Smith PM, McGregor A, Turnbull DM, Lightowlers RN. Synthesis of trifunctional PNA-benzophenone derivatives for mitochondrial targeting, selective DNA binding, and photo-cross-linking. *Bioconjug Chem.* 2003;14:962-966.

[175] Diviacco S, Rapozzi V, Xodo L, Helene C, Quadrifoglio F, Giovannangeli C. Site-directed inhibition of DNA replication by triple helix formation. *FASEB J.* 2001;15:2660-2668.

[176] Liebling MR, Jou NT, Fang W, Louie JS. Blockade of plasmid replication mediated by peptide nucleic acids *Mol Biotechnol.* 2003;25:229-240.

[177] Hirschman SZ, Chen CW. Peptide nucleic acids stimulate gamma interferon and inhibit the replication of the human immunodeficiency virus *J Investig Med.* 1996;44347-351.

[178] Arzumanov A, Walsh AP, Liu X, Rajwanshi VK, Wengel J, Gait MJ. Oligonucleotide analogue interference with the HIV-1 Tat protein-TAR RNA interaction. *Nucleosides Nucleotides Nucleic Acids.* 2001;20:471-480. .

[179] Arzumanov A, Walsh AP, Rajwanshi VK, Kumar R, Wengel J, Gait MJ. Inhibition of HIV-1 Tat-dependent trans activation by steric block chimeric 2'-O-methyl/LNA oligoribonucleotides. *Biochemistry* 2001; 40:14645-14654.

[180] Kaushik N, Basu A, Pandey VN. Inhibition of HIV-1 replication by anti-trans-activation responsive polyamide nucleotide analog. *Antiviral Res.* 2002;56:13-27.

[181] Kaushik N, Pandey VN. PNA targeting the PBS and A-loop sequences of HIV-1 genome destabilizes packaged tRNA3(Lys) in the virions and inhibits HIV-1 replication. *Virology.* 2002;303:297-308.

[182] Depecker G, Patino N, Di Giorgio C, Terreux R, Cabrol-Bass D, Bailly C, Aubertin AM, Condom R. Cyclic PNA-based compound directed against HIV-1 TAR RNA: modelling, liquid-phase synthesis and TAR binding. *Org Biomol Chem.* 2004;;2:74-79.

[183] Pesce CD, Bolacchi F, Bongiovanni B, Cisotta F, Capozzi M, Diviacco S, Quadrifoglio F, Mango R, Novelli G, Mossa G, Esposito C, Ombres D, Rocchi G, Bergamini A. Anti-gene peptide nucleic acid targeted to proviral HIV-1 DNA inhibits in vitro HIV-1 replication. *Antiviral Res.* 2005;66:13-22.

[184] Chaubey B, Tripathi S, Ganguly S, Harris D, Casale RA, Pandey VN A PNA-transportan conjugate targeted to the TAR region of the HIV-1 genome exhibits both antiviral and virucidal properties. *Virology.* 2005;331:418-428.

[185] Tripathi S, Chaubey B, Ganguly S, Harris D, Casale RA, Pandey VN. Anti-HIV-1 activity of anti-TAR polyamide nucleic acid conjugated with various membrane transducing peptides. *Nucleic Acids Res.* 2005; 33:4345-4356.

[186] Turner JJ, Ivanova GD, Verbeure B, Williams D, Arzumanov AA, Abes S, Lebleu B, Gait MJ. Cell-penetrating peptide conjugates of peptide nucleic acids (PNA) as inhibitors of HIV-1 Tat-dependent trans-activation in cells. *Nucleic Acids Res.* 2005;33:6837-6849.

[187] Turner JJ, Fabani M, Arzumanov AA, Ivanova G, Gait MJ. Targeting the HIV-1 RNA leader sequence with synthetic oligonucleotides and siRNA: Chemistry and cell delivery. *Biochim Biophys Acta.* 2005; Nov 18.

[188] Lansdorp, PM;, Verwoerd, N. P., van de Rijke, F. M., Dragowska, V., Little, M. T Dirks, R. W., Raap, A. K. and Tanke, H. J. (1996) Heterogeneity in telomere length of human chromosomes. *Hum. Mol. Genet.* 5, 685-691. 188a. Rufer N, Dragowska W, Thornbury G, Roosnek E, Lansdorp PM. Telomere length dynamics in human lymphocyte subpopulations measured by flow cytometry. *Nat Biotechnol.* 1998;16:0743-747.

[189] Slijepcevic P Telomere length measurement by Q-FISH *Methods in Cell Science* 2001;23:17–22.

[190] Williams B, Stender H, Coull JM. PNA fluorescent in situ hybridization for rapid microbiology and cytogenetic analysis. *Methods Mol Biol.* 2002;208:181-193.

[191] Schmid I, Dagarag MD, Hausner MA, Matud JL, Just T, Effros RB, Jamieson BD. Simultaneous flow cytometric analysis of two cell surface markers, telomere length, and DNA content. *Cytometry*. 2002;49:96-105.
[192] Shammas MA, Liu X, Gavory G, Raney KD, Balasubramanian S, Shmookler Reis RJ. Targeting the single-strand G-rich overhang of telomeres with PNA inhibits cell growth and induces apoptosis of human immortal cells. *Exp Cell Res*. 2004;295:204-214.
[193] O'Sullivan JN, Finley JC, Risques RA, Shen WT, Gollahon KA, Moskovitz AH, Gryaznov S, Harley CB, Rabinovitch PS. Telomere length assessment in tissue sections by quantitative FISH: image analysis algorithms. *Cytometry A*. 2004;58:120-131.
[194] Graakjaer J, Pascoe L, Der-Sarkissian H, Thomas G, Kolvraa S, Christensen K, Londono-Vallejo JA. The relative lengths of individual telomeres are defined in the zygote and strictly maintained during life. *Aging Cell*. 2004; 3:97-102.
[195] Arutyunyan R, Gebhart E, Hovhannisyan G, Greulich KO, Rapp A. Comet-FISH using peptide nucleic acid probes detects telomeric repeats in DNA damaged by bleomycin and mitomycin C proportional to general DNA damage. *Mutagenesis* 2004; 19:403-408.
[196] Hovhannisyan G, Rapp A, Arutyunyan R, Greulich KO, Gebhart E. Comet-assay in combination with PNA-FISH detects mutagen-induced DNA damage and specific repeat sequences in the damaged DNA of transformed cells. *Int J Mol Med*. 2005;15::437-442.
[197] Arutyunyan R, Rapp A, Greulich KO, Hovhannisyan G, Haroutiunian S, Gebhart E. Fragility of telomeres after bleomycin and cisplatin combined treatment measured in human leukocytes with the Comet-FISH technique. *Exp Oncol*. 2005 ;27:38-42.
[198] Derradji H, Bekaert S, Van Oostveldt P, Baatout S. Comparison of different protocols for telomere length estimation by combination of quantitative fluorescence in situ hybridization (Q-FISH) and flow cytometry in human cancer cell lines. *Anticancer Res*. 2005;25:1039-1050.
[199] Wick U, Kirsch M, Rauch A, Chudoba I, Lausen B, Efferth T, Gebhart E. FISH studies on the telomeric regions of the T-cell acute lymphoblastic leukemia cell line CCRF-CEM. *Cytogenet Genome Res*. 2005;111:34-40.
[200] Wick U, Gebhart E. The order of PNA-FISH-detected chromosomal telomere lengths in human T-cells is rather stable, even under the influence of strong mutagens. *Int J Mol Med*. 2005;16:951-957.
[201] Wick U, Gebhart E. Studies on the action of mitomycin C and bleomycin on telomere lengths of human lymphocyte chromosomes. *Int J Mol Med*. 2005;16:463-469.
[202] Piotrowska K, Kleideiter E, Murdter TE, Taetz S, Baldes C, Schaefer U, Lehr CM, Klotz U. Optimization of the TRAP assay to evaluate specificity of telomerase inhibitors Lab Invest. 2005 Dec;85:1565-1569.
[203] Potter AJ, Wener MH. Flow cytometric analysis of fluorescence in situ hybridization with dye dilution and DNA staining (flow-FISH-DDD) to determine telomere length dynamics in proliferating cells. *Cytometry A*. 2005: 68:53-58.
[204] Narath R, Lorch T, Greulich-Bode KM, Boukamp P, Ambros PF Automatic telomere length measurements in interphase nuclei by IQ-FISH. Cytometry A. 2005: 68:113-120.

[205] Pellestor F, Paulasova P. The peptide nucleic acids (PNAs): introduction to a new class of probes for chromosomal investigation. *Chromosoma*. 2004;112:375-380.

[206] Pellestor F, Paulasova P The peptide nucleic acids (PNAs), powerful tools for molecular genetics and cytogenetics. *Eur J Hum Genet*. 2004;12::694-700.

[207] Pellestor F, Paulasova P, Macek M, Hamamah S. The peptide nucleic acids: a new way for chromosomal investigation on isolated cells? *Hum Reprod*. 2004;19:1946-1951.

[208] Paulasova P, Pellestor F The peptide nucleic acids (PNAs): a new generation of probes for genetic and cytogenetic analyses. *Ann Genet*. 2004; 47:349-358.

[209] Pellestor F, Paulasova P, Macek M, Hamamah S. The use of peptide nucleic acids for in situ identification of human chromosomes *J Histochem Cytochem*. 2005;;53:395-400.

[210] Taneja KL. Localization of trinucleotide repeat sequences in myotonic dystrophy cells using a single fluorochrome-labeled PNA probe. *Biotechniques*. 1998; 24:472-476.

[211] Chen C, Hong YK, Ontiveros SD, Egholm M, Strauss WM. Single base discrimination of CENP-B repeats on mouse and human Chromosomes with PNA-FISH. *Mamm Genome*. 1999;10:13-18.

[212] Deng W, Lucas JN. Combined FISH with pan-telomeric PNA and whole chromosome-specific DNA probes to detect complete and incomplete chromosomal exchanges in human lymphocytes. *Int J Radiat Biol*. 1999;75:1107-1112.

[213] Chen C, Wu B, Wei T, Egholm M, Strauss WM. Unique chromosome identification and sequence-specific structural analysis with short PNA oligomers. *Mamm Genome*. 2000;11::384-391.

[214] Taneja KL, Chavez EA, Coull J, Lansdorp PM. Multicolor fluorescence in situ hybridization with peptide nucleic acid probes for enumeration of specific chromosomes in human cells. *Genes Chromosomes Cancer*. 2001;30:57-63.

[215] Pellestor F, Andreo B, Taneja K, Williams B. PNA on human sperm: a new approach for in situ aneuploidy estimation. *Eur J Hum Genet*. 2003;11:337-341.

[216] Paulasova P, Andreo B, Diblik J, Macek M, Pellestor F. The peptide nucleic acids as probes for chromosomal analysis: application to human oocytes, polar bodies and preimplantation embryos. *Mol Hum Reprod*. 2004;10:467-472.

[217] Pellestor F, Anahory T, Hamamah S. The chromosomal analysis of human oocytes. An overview of established procedures. *Hum Reprod Update*. 2005;11:15-32.

[218] . Codina-Pascual M, Oliver-Bonet M, Navarro J, Starke H, Liehr T, Gutierrez-Mateo C, Sanchez-Garcia JF, Arango O, Egozcue J, Benet J. FISH characterization of a dicentric Yq (p11.32) isochromosome in an azoospermic male. *Am J Med Genet A*. 2004;127:302-306.

[219] Agerholm IE, Ziebe S, Williams B, Berg C, Cruger DG, Petersen GB, Kolvraa S. Sequential FISH analysis using competitive displacement of labelled peptide nucleic acid probes for eight chromosomes in human blastomeres. *Hum Reprod*. 2005;20:1072-1077.

[220] Stender H, Lund K, Petersen KH, Rasmussen OF, Hongmanee P, Miorner H, Godtfredsen SE. Fluorescence In situ hybridization assay using peptide nucleic acid probes for differentiation between tuberculous and nontuberculous mycobacterium species in smears of mycobacterium cultures. *J Clin Microbiol*. 1999;37:2760-2765.

[221] Stender H, Mollerup TA, Lund K, Petersen KH, Hongmanee P, Godtfredsen SE. Direct detection and identification of Mycobacterium tuberculosis in smear-positive sputum samples by fluorescence in situ hybridization (FISH) using peptide nucleic acid (PNA) probes. *Int J Tuberc Lung Dis.* 1999;3:830-837.

[222] Drobniewski FA, More PG, Harris GS. Differentiation of Mycobacterium tuberculosis complex and nontuberculous mycobacterial liquid cultures by using peptide nucleic acid-fluorescence In situ hybridization probes. *J Clin Microbiol.* 2000;38:444-447.

[223] Perry-O'Keefe H, Rigby S, Oliveira K, Sorensen D, Stender H, Coull J, Hyldig-Nielsen JJ. Identification of indicator microorganisms using a standardized PNA FISH method. *J Microbiol Methods.* 2001;47:281-292.

[224] Williams B, Stender H, Coull JM. PNA fluorescent in situ hybridization for rapid microbiology and cytogenetic analysis. *Methods Mol Biol.* 2002;208:181-93.

[225] Oliveira K, Brecher SM, Durbin A, Shapiro DS, Schwartz DR, De Girolami PC, Dakos J, Procop GW, Wilson D, Hanna CS, Haase G, Peltroche-Llacsahuanga H, Chapin KC, Musgnug MC, Levi MH, Shoemaker C, Stender H. Direct identification of Staphylococcus aureus from positive blood culture bottles. *J Clin Microbiol.* 2003;41:889-891.

[226] Chapin K, Musgnug M. Evaluation of three rapid methods for the direct identification of Staphylococcus aureus from positive blood cultures. *J Clin Microbiol.* 2003;41:4324-4327.

[227] Gonzalez V, Padilla E, Gimenez M, Vilaplana C, Perez A, Fernandez G, Quesada MD, Pallares MA, Ausina V. Rapid diagnosis of Staphylococcus aureus bacteremia using S. aureus PNA FISH. *Eur J Clin Microbiol Infect Dis.* 2004; 23:396-398.

[228] Hartmann H, Stender H, Schafer A, Autenrieth IB, Kempf VA. Rapid identification of Staphylococcus aureus in blood cultures by a combination of fluorescence in situ hybridization using peptide nucleic acid probes and flow cytometry. *J Clin Microbiol.* 2005;43:4855-4857.

[229] Xi C, Balberg M, Boppart SA, Raskin L. Use of DNA and peptide nucleic acid molecular beacons for detection and quantification of rRNA in solution and in whole cells. *Appl Environ Microbiol.* 2003;69:5673-5678.

[230] Lehtola MJ, Loades CJ, Keevil CW. Advantages of peptide nucleic acid oligonucleotides for sensitive site directed 16S rRNA fluorescence in situ hybridization (FISH) detection of Campylobacter jejuni, Campylobacter coli and Campylobacter lari. *J Microbiol Methods.* 2005;62:211-219.

[231] Zhdanova NS, Karamisheva TV, Minina J, Astakhova NM, Lansdorp P, Kammori M, Rubtsov NB, Searle JB. Unusual distribution pattern of telomeric repeats in the shrews Sorex araneus and Sorex granarius. *Chromosome Res.* 2005;13:617-625.

[232] Chen M, Liu M, Yu L, Cai G, Chen Q, Wu R, Wang F, Zhang B, Jiang T, Fu W. Construction of a novel peptide nucleic acid piezoelectric gene sensor microarray detection system. *J Nanosci Nanotechnol.* 2005;5:1266-1272.

[233] Malchere C, Verheijen J, van der Laan S, Bastide L, van Boom J, Lebleu B, Robbins I A short phosphodiester window is sufficient to direct RNase H-dependent RNA cleavage by antisense peptide nucleic acid. *Antisense Nucleic Acid Drug Dev.* 2000;10:463-468.

[234] Nielsen, P. E., Egholm, M. Berg, R. H: and Buchardt, O. (1993) Peptide nucleic acids (PNAs): potential antisense and antigene agents *Anti-Cancer Drug Des.* 8, 53-63.

[235] Lutz, M. J., Benner, S. A., Hein, S., Breipohl, G. and Uhlmann, E. (1997) Recognition of Uncharged Polyamide-Linked Nucleic Acid Analogs by DNA Polymerases and Reverse Transcriptases. *J. Am. Chem. Soc.* 119, 3177-3178.

[236] Koppelhus, U., Zachar, V., Nielsen, P. E., Liu, X., Eugen-Olsen, J. and Ebbesen, P. (1997) Efficient in vitro inhibition of HIV-1 gag reverse transcription by peptide nucleic acid (PNA) at minimal ratios of PNA/RNA. *Nucleic Acids Res.* 25, 2167-2173.

[237] Norton, J. C., Piatyszek, M. A., Wright, W. E., Shay, J. W. and Corey, D. R. (1996) Inhibition of human telomerase activity by peptide nucleic acids. *Nat. Biotech.* 14, 615-620.

[238] Hamilton, S. E., Simmons, C. G., Kathiriya, I. S. and Corey, D. R. (1999) Cellular delivery of peptide nucleic acids and inhibition of human telomerase. *Chem. Biol.* 6, 343-351.

[239] Faruqi, A. F., Egolm, M. and Glazer, P. M. (1997). Peptide nucleic acid-targeted mutagenesis of a chromosomal gene in mouse cells. *Proc. Natl. Acad Sci. U.S.A.* 95, 1398-1403.

[240] Pooga, M. et. al and Langel, U. (1998) Cell penetrating PNA constructs regulate galanin receptor levels and modify pain transmision *in vivo*. *Nat. Biotechnol.* 16, 857-861.

[241] Basu, S. and Wickstrom, E. (1997) Synthesis and characterization of a peptide nucleic acid conjugated to a D-peptide analog of insulin-like growth factor 1 for increased cellular uptake. *Bioconj. Chem.* 8, 481-488.

[242] Simmons, C. G., Pitts, A. E., Mayfield, L. D., Shay, J. W. and Corey, D. R. (1997) Synthesis and permeability of PNA-peptide conjugates. *Bioorg. Med. Chem. Lett.* 7, 3001-3007.

[243] Noble, S. A., Bonham, M. A., Bisi, J. E., Bruckenstein, D. A., Brown, P. H., Brown, S. C., Cadilla, R., Gaul, M. D., Hanvey, J. C., Hassman, C. F., Josey, J. A., Luzzio, M. J., Myers, P. M., Pipe, A. J., Ricca, D. J., Su, C. W., Stevenson, C. L., Thomson, S. A., Wiethe, R. W. and Babiss, L. E. (1995) Impact of biophysical parameters on the biological assessment of peptide nucleic acids, antisense inhibitors of gene expression. *Drug Dev. Res.* 34, 184-195.

[244] Uhlmann, E. Will, D. W., Breiphol, G., Langer, D. and Ryte, A. (1996) *Angew. Chem.* 108, 2793-2797.

[245] Uhlmann, E. Will, D. W., Breiphol, G., Langer, D. and Ryte, A. (1996) Synthesis and properties of PNA/DNA chimeras. *Angew. Chem. Int. Ed. Engl.* 35, 2632-2635.

[246] Aldrian-Herrada, G., Desarme´nien, M. G., Orcel, H., Boissin-Agasse, L., Me´ry, J., Brugidou, J. and Rabie´ A. (1998) A peptide nucleic acid (PNA) is more rapidly internalized in culture neurons when coupled to a retro-inverso delivery peptide. The antisense activity depresses the target mRNA and protein in magnocellular oxytocin neurons. *Nucleic Acids Res.* 26, 4910-4916.

[247] Krotz, A. H., Buchardt, O. and Nielsen, P. E. (1995) Synthesis of 'Retro-Inverso' peptide nucleic acids: 1. Characterization of the monomers. *Tetrahedron Letters*, 36, 6937-6940.

[248] Krotz, A. H., Buchardt, O. and Nielsen, P. E. (1995) Synthesis of 'Retro-Inverso' peptide nucleic acids: 2. Oligomerization and stability. *Tetrahedron Letters*, 36, 6941-6944.

[249] Gallazzi F, Wang Y, Jia F, Shenoy N, Landon LA, Hannink M, Lever SZ, Lewis MR. Synthesis of radiometal-labeled and fluorescent cell-permeating peptide-PNA conjugates for targeting the bcl-2 proto-oncogene. *Bioconjug Chem.* 2003;14:1083-1095.

[250] Gait MJ. Peptide-mediated cellular delivery of antisense oligonucleotides and their analogues. *Cell Mol Life Sci.* 2003; 60:844-853.

[251] Fisher L, Soomets U, Cortes Toro V, Chilton L, Jiang Y, Langel U, Iverfeldt K. Cellular delivery of a double-stranded oligonucleotide NFkappaB decoy by hybridization to complementary PNA linked to a cell-penetrating peptide. *Gene Ther.* 2004;11:1264-1272.

[252] Diaz-Mochon JJ, Bialy L, Watson J, Sanchez-Martin RM, Bradley M. Synthesis and cellular uptake of cell delivering PNA-peptide conjugates. *Chem Commun (Camb).* 2005;4::3316-3318.

[253] Kilk K, Langel U. Cellular delivery of peptide nucleic acid by cell-penetrating peptides. *Methods Mol Biol.* 2005;298:131-141.

[254] Doyle DF, Braasch DA, Simmons CG, Janowski BA, Corey DR. Inhibition of gene expression inside cells by peptide nucleic acids: effect of mRNA target sequence, mismatched bases, and PNA length. *Biochemistry.* 2001;40:53-64.

[255] Yuan X, Ma Z, Zhou W, Niidome T, Alber S, Huang L, Watkins S, Li S. Lipid-mediated delivery of peptide nucleic acids to pulmonary endothelium. Biochem Biophys Res Commun. 2003;302:6-11.

[256] Boado RJ, Tsukamoto H, Pardridge WM. Drug delivery of antisense molecules to the brain for treatment of Alzheimer's disease and cerebral AIDS. *J Pharm Sci.* 1998; 87:1308-1315.

[257] Koppelhus U, Nielsen PE. Cellular delivery of peptide nucleic acid (PNA). *Adv Drug Deliv Rev.* 2003;55:267-280.

[258] Ardhammar, M., Nordén, B., Nielsen, P. E., Malmström, B. G. and Wittung-Stafshede, P. *In vitro* membrane of modified Peptide Nucleic Acid (PNA). *J. Bio. Struct. Dyn.* (1999); 17:33-40. 258b. Wittung, P., Kajanus, J., Edwards, K., Nielsen, P., Nordén, B. and Malmström, B.G. Phospholipid membrane permeability of peptid nucleic acid. *FEBS Letters* 1995; 365:27-29.

[259] Ljungstrom T, Knudsen H, Nielsen PE. Cellular uptake of adamantyl conjugated peptide nucleic acids *Bioconjug Chem.* 1999;10::965-972.

[260] Mologni L, Marchesi E, Nielsen PE, Gambacorti-Passerini C. Inhibition of promyelocytic leukemia (PML)/retinoic acid receptor-alpha and PML expression in acute promyelocytic leukemia cells by anti-PML peptide nucleic acid. *Cancer Res.* 2001;61:5468-5473.

[261] Nastruzzi C, Cortesi R, Esposito E, Gambari R, Borgatti M, Bianchi N, Feriotto G, Mischiati C. Liposomes as carriers for DNA-PNA hybrids. *J Control Release.* 2000;68:237-249.

[262] Borgatti M, Breda L, Cortesi R, Nastruzzi C, Romanelli A, Saviano M, Bianchi N, Mischiati C, Pedone C, Gambari R. Cationic liposomes as delivery systems for double-stranded PNA-DNA chimeras exhibiting decoy activity against NF-kappaB transcription factors. *Biochem Pharmacol.* 2002;64:609-616.

[263] Borgatti M, Romanelli A, Saviano M, Pedone C, Lampronti I, Breda L, Nastruzzi C, Bianchi N, Mischiati C, Gambari R. Resistance of decoy PNA-DNA chimeras to enzymatic degradation in cellular extracts and serum. *Oncol Res.* 2003;13:279-287.

[264] Ratilainen, T., Lincoln, P. and Nordén, B. A simple model for gene-targeting *Biophys.J.* 2001;81:2876-2885.

[265] Lagriffoule, P. Wittung, P. Eriksson, M. Jensen, K. K. Nordén, Buchardt, O. and Nielsen, P. E. Peptide Nucleic Acids with a conformationally constrained chiral cyclohexyl-derived backbone. *Chem. Eur. J.* 1997; 3:912-919.

[266] Efimov, V. A., Choob, M. V., Buryakova, A. A., Kalinkina, A. L., and Chakhmakhcheva, O. G. (1998) Synthesis and Evaluation of some Properties of Chimeric Oligomers containing PNA and Phosphono-PNA Residues. *Nucleic Acids Res.* 26, 566-575.

[267] Armitage, B., Ly, D., Koch, T., Frydenlund, H., Örum, H., and Schuster, G. B. (1998) Hairpin.Forming Peptide Nucleic Acid Oligomers. *Biochemistry* 37, 9417-9425.

[268] Chirstensen, L., Hansen, H. F., Koch, T. and Nielsen, P. E. (1998) Inhibition of PNA Triplex Formation by N^4-benzoylated Cytosine. *Nucleic Acids Res.* 26, 2735-2739.

[269] Nielsen, P. E. (1999) Applications of peptide nucleic acids. *Curr. Opin. Biotechnol.* 10, 71-75.

[270] Hansen, H. F., Christensen, L., Dahl, O and Nielsen, P. E. (1999) 6-thioguanine in peptide nucleic acids: synthesis and hybridization properties. *Nucleosides Nucleotides* 18, 5-9.

[271] Falkiewicz, B., Kowalska, K., Kolodziejczyk, A. S., Wisniewski, K. and Lankiewicz, L. (1999) Synthesis of new chiral peptide nucleic acid (PNA) monomers by a simplified reductive animation method. *Nucleosides Nucleotides* 18, 353-361.

[272] Lutz, M. J., Will, D. W., Breiphol, G., Brenner, S. A. and Uhlmann, E. (1999) Synthesis of a monocharged peptide nucleic acid (PNA) analogu and its recognition as substrate by DNA polymerases. *Nucleosides Nucleotides* 18, 393-40.

[273] Zatsepin TS, Turner JJ, Oretskaya TS, Gait MJ. Conjugates of oligonucleotides and analogues with cell penetrating peptides as gene silencing agents *Curr Pharm Des.* 2005;11:3639-3654.

[274] Porcheddu A, Giacomelli G. Peptide nucleic acids (PNAs), a chemical overview *Curr Med Chem.* 2005;12:2561-2599.

[275] Zhilina ZV, Ziemba AJ, Ebbinghaus SW. Peptide nucleic acid conjugates: synthesis, properties and applications. *Curr Top Med Chem.* 2005;5:1119-1131.

[276] Hudson RH, Wojciechowski Synthesis and properties of PNA oligomers containing orotic acid derivatives *Nucleosides Nucleotides Nucleic Acids.* 2005;24:1123-1126.

[277] Tian X, Aruva MR, Wolfe HR, Qin W, Sauter ER, Thakur ML, Waldman SA, Wickstrom E. Tumor-targeting peptide-PNA-peptide chimeras for imaging overexpressed oncogene mRNAs *Nucleosides Nucleotides Nucleic Acids.* 2005;24:1085-1091.

[278] Dose C, Seitz O. Convergent synthesis of peptide nucleic acids by native chemical ligation. *Org Lett.* 2005;7:4365-4368.
[279] Ma LJ, Zhang GL, Chen SY, Wu B, You JS, Xia CQ, Yu XQ. The first synthesis of chiral PNA monomer-cyclen conjugates. *J Pept Sci.* 2005;11:812-817.

In: PRINS and PNA Technologies in Chromosomal Investigations ISBN 1-60021-413-4
Editor: Franck Pellestor, pp. 143-168

Chapter 10

PNA Suppression Method Combined with Fluorescence *In Situ* Hybridisation (FISH) Technique

Kirsten Vang Nielsen*, Sven Müller, Inge E. Agerholm**, Tim Svenstrup Poulsen, Steen Hauge Matthiesen, Caroline Janet Witton and Andreas Schønau
Dako, Glostrup and Brædstrup Sygehus**, Denmark

Abstract

The successful use of large genomic probes for fluorescence *in situ* hybridisation (FISH) is dependent on blocking of the undesired background staining derived from repetitive sequences that are present throughout the genome. Traditionally, this has been achieved by using a component of the total DNA, Cot-1 DNA, enriched with repetitive sequences. Within and around the genes, the most frequent repetitive element is the Alu sequences. By selection of specific PNA oligos, directed towards the Alu sequences, an effective suppression of the background has been achieved. The selected Alu PNA Mixture can be used in protocols with a variety of different probes. Also, it can be used in combination with labelled PNA probes targeting other repetitive sequences in the genomes, e.g. the centromeres. The suppression method can be used to analyse cells for the occurrence of chromosomes, chromosome fragments, genes, or chromosome aberrations (e.g. translocations, deletions, amplifications) associated with a condition or disease. Any method that can detect, identify and/or quantify selected target genomic nucleic acid in metaphase spreads, interphase nuclei, tissue sections, and extracted DNA from these cells can potentially take advantage of the PNA blocking approach. These methods include CISH (chromogenic *in situ* hybridisation), FISH, multi-colour FISH, Fiber-FISH, Comparative Genomic Hybridisation (CGH), chromosome paints and the analysis of BAC clones and arrays. Another way of using PNA for suppression is based

* Correspondence to: Dr. Kirsten Vang Nielsen, Molecular Pathology, Dako Denmark A/S, Produktionsvej 42, DK-2600 Glostrup, Denmark. Tel: +45 44 85 94 94; Email: kirsten.vang@dako.com

on the sequential FISH analysis with PNA oligos and competitive displacement of the previous bound PNAs using mixture of labelled and unlabelled PNA oligos. This method is especially an advantage when only limited sample is available for multiple analyses as in preimplantation genetic diagnosis.

Keywords: Alu sequences, FISH, PNA

Introduction

Fluorescence *in situ* hybridisation (FISH) has become an important tool for determining the number, size and/or location of specific DNA sequences in mammalian cells. DNA sequences ranging from whole genomes down to several kilobases can be studied using current hybridisation techniques based on fluorescence microscopy. In Comparative Genomic Hybridisation (CGH) the entire genome is stained and compared to a normal reference genome for the detection of regions with aberrant copy number [1]. In the m-FISH technique (multi colour FISH) each normal chromosome is stained by a separate colour [2]. When used on abnormal material, the probes will stain the aberrant chromosomes thereby deducing the normal chromosomes from which they are derived [3]. Specific DNA sequences, such as the ABL gene, can be reliably stained using probes of only 15 kb [4]. FISH-based staining is sufficiently distinct to allow detection of hybridisation signals in both metaphase spreads, interphase nuclei, and in cells from tissue sections. Single and multicolour FISH, using nucleic acid probes, have been applied to different clinical applications generally known as molecular cytogenetics, including prenatal diagnosis, leukaemia diagnosis, and tumour cytogenetics.

Repetitive Sequences

A large component of the human genome is comprised of repeated sequences. Heat denaturation and reannealing studies on DNA of higher organisms have distinguished three populations of eukaryotic DNA; a quickly reannealing component representing 25% of the total DNA, an intermediate component that represents 30% of the total DNA, and a slow component that represents 45% of the total DNA [5]. Sequence analysis has shown that the slow component is made up by unique sequences, which include the protein encoding genes, while the fast and intermediate components represent repetitive sequences. The fast component contains small (a few nucleotides long), highly repetitive DNA sequences, which are usually found in tandem while the intermediate component contains the interspersed repetitive DNA [6,7]. The repetitive units of the intermediate component are interspersed within the genome and are the major reason that large genomic nucleic acid probes are not well suited for hybridisation analysis without blocking of the undesired staining derived from the repetitive elements.

Interspersed repeated sequences are classified as either SINEs (short interspersed repeated sequences) or LINEs (long IRSs) [8,9]. In primates, each of these classes is dominated by a single DNA sequence family, both of which are classified as retroposons [10]. The Alu repeat DNA sequence family makes up the majority of human SINEs. The Alu repeat DNA family members are characterized by a consensus sequence of approximately 280 to 300 bp that consist of two similar sequences arranged as a head to tail dimer. Approximately one million copies of the Alu repeat sequence have been estimated to be present per haploid human genome, thereby representing about ten percent of the genome [11]. This estimate showed to be consistent with the sequence determination of the human genome [12] reporting 10.60% Alu repeats. However, Alu repeats are not evenly distributed in the genome. Chromosome 21 and 22 contain 9.48% and 16.80% Alu repeats, respectively [12,13].

Alu elements have amplified in the genome of primates through retroposition over the past 65 million years and have been organized into a wealth of overlapping subfamilies based on diagnostic mutations shared by subfamily members [14]. Batzer et al. described a consensus nomenclature for Alu repeats sequences; representing the oldest (J), intermediate (S) and young (Y) family branches. Only the Y family branch is still transcriptional active but it is very small as each of the three subfamily members have produced less than 2000 elements [15]. It has been calculated that of the primate Alu repeat family branches, approximately one-fifth belong to the J family and four-fifths to the S family [16]. The S family is dominated by the Sx subfamily as it represents more than 50% of the total S family branch.

In addition to SINEs and LINEs, there are several other types of repeats that are known to exist in the genome of humans as well as in other organisms. Chromosome telomeres are repeat sequences that appear to exist only at the termini of all chromosomes. They are believed to shorten during the life of an organism and may play a role in the aging of an organism [17]. Likewise, chromosome centromeres contain distinct repeat sequences that exist only in the central region of a chromosome. Certain of the centromere repeat sequences can be detected in all chromosomes of an organism whilst other repeat sequences are unique to a particular chromosome and can be used to identify specific chromosomes [18]. Both telomeres and centromeres can be visualized using labelled PNA probes, as well as other targets present in high copy number [19-21]. Telomere and centromere repeat sequences differ from interspersed repetitive sequence, such as SINEs and LINEs, in that the telomere and centromere repeat sequences are localized within a certain region of the chromosome. By comparison, SINEs and LINEs are dispersed randomly throughout the entire genome [22].

The Need for Blocking Techniques in FISH Analysis

Detection of specific nucleic acid sequences by *in situ* hybridisation using non-radioactive labels has been applied for almost twenty years. However, the randomly distributed repeat sequences, especially SINEs, are problematic for the use of specific probes, derived from large genomic clones, because the probes are likely to contain randomly distributed repeat sequence. Consequently, the labelled probes hybridise not only to the desired target, but also to the repeat sequences randomly distributed in the genome, thereby generating unspecific background.

Refinement of non-radioactive detection and visualization methods has resulted in improved detection limits and thereby allowed the localization of large single-copy sequences [23]. Historically, in the first study it was necessary to construct a mixture of seven subclones (a total of 22.3 kb derived from a cosmid DNA clone containing the 3' end of the *Tg* gene) in order to eliminate highly repeated sequences present in the original genomic cosmid DNA. Although this was an improvement, a more attractive strategy, based on direct use of large genomic cloned segments in combination with Cot-1 DNA, was soon developed. The use of Cot-1 DNA eliminates background signal caused by highly repetitive sequences, by introducing a competitive hybridisation process [24].

Cot-1 DNA is a heterogeneous mixture that is prepared by degrading total human DNA and processing the resulting material to select for those fragments that are enriched in the repeated sequences [25]. Although the use of Cot-1 DNA has proven to be effective in suppressing undesired binding of the repetitive sequences, there are several disadvantages to this method. One such disadvantage pertains to the preparation of the Cot-1 DNA itself. The starting material itself is not well defined and is likely to vary from sample to sample. Furthermore, the processing methods will produce material that varies from batch to batch. Additionally, the Cot-1 DNA is a heterogeneous mixture of fragments that is impossible to completely characterize and define. Hence, the batch-to-batch variability, as well as the inability to characterize the Cot-1 DNA product, leaves substantial room for improvement [26,27].

The PNAs can be chemically synthesized and purified in a manner that provides for proper characterization. Consequently, the unlabelled PNAs can be individually prepared, characterized and quantitated before preparing a mixture of PNA oligos for a blocking application. Hence, the mixture of PNA oligos itself can therefore be more carefully controlled, characterized and reproduced than the Cot-1 DNA. Since one possible application for such a mixture of PNA oligos is in a diagnostic assay, the ability to more easily characterize and reproduce the exact composition of the mixture from one batch to the next is advantageous.

Methods

Protocol for PNA/DNA FISH on Cytological Specimens

The protocol is intended for use on cytological specimens, such as e.g. metaphase spreads, blood smears, cell lines etc. The reagents used in this protocol are from Dako's Cytology FISH Accessory Kit. The kit contains all the key reagents, except for probe, required to complete a FISH procedure for cytological specimens.

Immerse slides containing specimen in Wash Buffer with 3.7% formaldehyde for 2 min. Wash in Wash Buffer 2 x 5 min, dehydrate in an ethanol series (70%, 85%, and 96% ethanol) for 2 min each and air-dry. Apply 10 µL DNA/PNA probe on target area, followed by e.g. an 18x18 mm coverslip. Seal the edges of the coverslip carefully with plenty of Coverslip

Sealant and let slides air-dry for a few minutes until the Sealant has set. Denature at 82°C for 5 minutes and hybridise over night at 45°C on a Hybridizer (Dako) or equivalent. Note that to obtain good reproducible results a high controlled relative humidity as well as controlled temperature is important during the hybridisation step.

After the hybridisation step, remove the coverslips and immerse slides in Stringency Buffer at room temperature (RT, 20-25°C). Transfer slides to 65°C pre-heated Stringency Buffer for 10 min. Wash slides in Wash Buffer 2 x 3 min, dehydrate in an ethanol series (70%, 85%, and 96% ethanol) for 2 min each and air-dry. Finally, mount slides with 10-15 μL Fluorescence Mounting Medium with blue counter stain and seal with a coverslip. Keep slides in dark for a minimum of 15 min before reading on a fluorescence microscope.

Protocol for PNA/DNA FISH on Formalin-fixed Paraffin-embedded Tissue Sections

The most common way to fix the tissue, before the paraffin embedding, is with neutral buffered formalin. Other fixative may be used, but the quality of the end result might be reduced. The described protocol is based on the use of formalin as fixative. Properly fixed and embedded tissues will keep indefinitely prior to sectioning and slide mounting if stored correctly (15-25°C) [28,29].

Cut formalin-fixed paraffin embedded tissue specimens into sections of 2-6 μm, using water bath collect sections on slides (Silanized Slides, Dako; or poly-L-Lysine coated slides) and then air-dry. Melt the paraffin at 60°C for 30-60 min. Generally, the thinner sections produce better FISH performance.

The reagents used in this protocol are from Dako's Histology FISH Accessory Kit. The kit contains all the key reagents, except for probe, required to complete a FISH procedure for formalin-fixed, paraffin-embedded tissue section specimens.

The paraffin-embedded tissues sections are deparaffinated in xylene baths 2 x 5 min, rehydrated in 96% ethanol baths 2 x 2 min, followed by 70% ethanol baths 2 x 2 min and then transferred to Wash Buffer. A xylene substitute can be used instead of xylene.

Immerse the deparaffinated slides in Pre-Treatment Solution and incubate at minimum 95°C for 10 min. Remove the container holding the slides and allow to cool for 15 min at RT. Wash slides in Wash Buffer 2 x 3 min. Note that the closer the Pre-Treatment Solution is to 100°C the better the FISH signals obtained. A water bath can be used for the pre-treatment step or alternatively a microwave oven capable of controlling temperature. A controlled use of microwave oven is a fast, convenient and excellent alternative to a water bath. A pressure cooker like the Pascal (Dako) can be used as well, 121°C for 1 min, e.g. for laboratories placed at a high altitude that thereby having difficulties in obtaining a sufficient high temperature. The optimum temperature and time might be different for other pressure cookers.

Drip enough Ready-to-Use Pepsin to cover sections and incubate at 37°C on Hybridizer (Dako) or on a heat plate for 3-6 minutes. Note that in order to control the digestion the heat plate should be stable and validated. A set incubation time of 3-6 minutes will be adequate for most sections. It is important to use a correct digestion time and as such the time required might need to be optimised to fit to local fixation protocols. Use of different fixation methods locally or country wise can increase the suggested digestion time to 7-15 min or even more. To determine the local optimum digestion time it is therefore recommended to digest four sections cut from the same tissue block for 3, 6, 12, and 18 minutes, respectively.

Wash slides in Wash Buffer 2 x 3 min, dehydrate in an ethanol series (70%, 85%, and 96% ethanol) for 2 min each and air-dry.

For the hybridisation step apply 10 μL DNA/PNA probe to the centre of the tissue section of the dehydrated slides. Apply e.g. a 22x22 mm coverslip. Seal the edges of the coverslip carefully with plenty of Coverslip Sealant and air-dry for a few minutes until the Sealant has set. Denature at 82°C for 5 min and hybridise over night at 45°C on a Hybridizer or equivalent. Note that to obtain good reproducible results a high controlled relative humidity as well as a controlled temperature is important during the hybridisation step.

After the hybridisation step, remove the coverslips and immerse the slides in Stringency Buffer at RT. Transfer slides to pre-heated Stringency Buffer (65°C) for 10 min. Wash slides in Wash Buffer 2 x 3 min, dehydrate sections in an ethanol series (70%, 85%, and 96% ethanol) for 2 min each and air-dry. Finally, mount slides with 10-15 μL Fluorescence Mounting Medium with blue counter stain and seal with a coverslip. Keep slides in dark for a minimum of 15 min before reading on a fluorescence microscope. The signal intensity improves by prolonging the time between mounting and reading to e.g. 1-2 h.

Protocol for Sequential FISH Using Competitive Displacement

Interphase nuclei and metaphase spreads were obtained from peripheral blood lymphocytes of individuals with normal karyotype. These experiments were carried out on lymphocyte nuclei and metaphases with either probe set A or B [30]. After inspection and measurement of signals, a subsequent FISH cycle was performed where the slides were denatured at either 20° C, 30° C, 40° C, 50° C, 60° C and 70° C in the presence or absence of unlabelled probe with same composition as in the first FISH cycle. After the procedure nuclei and metaphases were re-inspected and signals measured again in order to estimate the amount of labelled probe from the first cycle that was displaced at each temperature.

Fluorescence Microscopy and Digital Imaging

Results were evaluated using a Leica fluorescence microscope (Leica Imaging System Ltd, Cambridge, UK) equipped with a 100 x oil immersion objective, a 10 x ocular (total enlargement is 1,000 fold). Filter cubes were either single bandpass filters for DAPI (A, P31000, Chroma Technology Corp., Rockingham, Vermont, US), FITC (L5, 513849, Chroma) and Texas Red (HQ-TRITC, 41002B, Chroma) or FITC/Texas Red dual bandpass filters (XF-53, Omega Optical Inc. Brattleboro, Vermont, US or P51006, Chroma). Electronic digital images were made of the slide using a Photometric Sensys CCD-camera and Leica QFISH software (Leica Imaging System Ltd, Cambridge, UK). Sequential FISH signal intensity was measured using the software from IPLAB (Scanalytics, Virginia, USA).

Results and Discussion

Suppression of Background Staining Using Alu PNAs

When designing a PNA blocking mix, it was envisioned that the PNA sequences that would be best for suppressing the undesired signal from genomic probes, would be those targeting the most prevalent Alu repeat sequences. Hence, a consensus sequence of known Alu repeat sequences was generated by performing an alignment analysis of five Alu consensus sequences representing the two family branches J and S (GenBank Acc. No. U14567 and U14571-14574) [31], using the Clustal W algorithm. The three identified subfamilies of the Y family branch have not been included in the alignment shown in Figure 1 as the relative frequency of Alu elements belonging to the Y family is very low, presumably less than 0.5% [15].

From the consensus sequence data, the sequence of numerous potential blocking PNAs was determined. The sequence of the various PNAs was selected to cover the upper and lower strand of the Alu consensus sequence (Figure 2), in order to disrupt as much as possible the hybridisation between the individual strands of the repeated Alu sequences of the genomic nucleic acid of a sample. It was preferable to cover at least fifty percent of the linear double stranded molecule to thereby prevent the rehybridisation of the two strands of genomic nucleic acid. Additionally, the PNA oligo candidates were checked for oligo self-dimerization, oligo pair dimers, and oligo hairpin loops to minimize the hybridisation reactivity with the probes themselves and with other PNAs in the mixture (i.e. intramolecular and intermolecular interactions, respectively). Using these design parameters, sequences were chosen for numerous PNAs that might, when mixed together, suppress the binding of labelled DNA probes to undesired target genomic nucleic acid. These sequences were used to prepare PNAs for further testing and analysis.

Evaluation of PNA Oligo Candidates

On metaphase spreads, hybridisation of labelled nucleic acid, containing Alu sequences, will produce a distinct and highly reproducible R-banding pattern [9,32]. Thus, it seemed reasonable to prepare labelled PNA probes having the various chosen sequences and then determine whether or not they would produce similar R-banding patterns as a way to test their binding to the Alu repeat sequences and thereby score the result for each PNA. Particularly because it was a consensus sequence that was being used to produce the PNA oligos, and not one of the actual naturally occurring Alu repeat sequences, testing was believed to be necessary to determine which of the PNAs that hybridised most strongly to the Alu repeat sequences of genomic nucleic acid.

```
                      --------GGCCGGGCGCGGTGGCTCACGCCTGTAATCCCAGCACTTTGG
                              10        20        30        40        50
HSU14574.seq Sx       CGGGCGGAGGCCGGGCGCGGTGGCTCACGCCTGTAATCCCAGCACTTTGG   50
HSU14573.seq Sq       --------GGCCGGGCGCGGTGGCTCACGCCTGTAATCCCAGCACTTTGG   42
HSU14572.seq Sp       --------GGCCGGGCGCGGTGGCTCACGCCTGTAATCCCAGCACTTTGG   42
HSU14571.seq Sc       --------GGCCGGGCGCGGTGGCTCACGCCTGTAATCCCAGCACTTTGG   42
HSU14567.seq J        --------GGCCGGGCGCGGTGGCTCACGCCTGTAATCCCAGCACTTTGG   42

                      GAGGCCGAGGCGGGCGGATCACCTGAGGTCAGGAGTTCGAGACCAGCCTG
                              60        70        80        90       100
HSU14574.seq Sx       GAGG--AAG---------ATCACCTGAGGTCAGGAGTTCGAGACCAGCCTG   90
HSU14573.seq Sq       GAGGCCGAGGCGGGTGGATCACCTGAGGTCAGGAGTTCGAGACCAGCCTG   92
HSU14572.seq Sp       GAGGCCGAGGCGGGCGGATCACCTGAGGTCGGGAGTTCGAGACCAGCCTG   92
HSU14571.seq Sc       GAGGCCGAGGCGGGCGGATCAC--GAGGTCAAGAGATCGAGACCATCCTG   90
HSU14567.seq J        GAGGCCGAGGCGGGAGGATCACTTGAGCCCAGGAGTTCGAGACCAGCCTG   92

                      GCCAACATGGTGAAACCCCGTCTCTACTAAAAATACAAAAATTAGCCGGG
                             110       120       130       140       150
HSU14574.seq Sx       GCCAACATGGTGAAACCCCGTCTCTACTAAAAATACAAAAATTAGCCGGG  140
HSU14573.seq Sq       GCCAACATGGTGAAACCCCGTCTCTACTAAAAATACAAAAATTAGCCGGG  142
HSU14572.seq Sp       ACCAACATGGAGAAACCCCGTCTCTACTAAAAATACAAAAATTAGCCGGG  142
HSU14571.seq Sc       GCCAACATGGTGAAACCCCGTCTCTACTAAAAATACAAAAATTAGCTGGG  140
HSU14567.seq J        GGCAACATAGTGAAACCCCGTCTCTACAAAAAATACAAAAATTAGCCGGG  142

                      CGTGGTGGCGCGCCTGTAATCCCAGCTACTCGGGAGGCTGAGGCAGGA
                             160       170       180       190       200
HSU14574.seq Sx       CGTGGTGGCGCGCCTGTAATCCCAGCTACTCGGGAGGCTGAGGCAGGA  190
HSU14573.seq Sq       CGTGGTGGCGGGCGCCTGTAATCCCAGCTACTCGGGAGGCTGAGGCAGGA  192
HSU14572.seq Sp       CGTGGTGGCGCATGCCTGTAATCCCAGCTACTCGGGAGGCTGAGGCAGGA  192
HSU14571.seq Sc       CGTGGTGGCGCGCGCCTGTAGTCCCAGCTACTCGGGAGGCTGAGGCAGGA  190
HSU14567.seq J        CGTGGTGGCGCGCGCCTGTAGTCCCAGCTACTCGGGAGGCTGAGGCAGGA  192

                      GAATCGCTTGAACCCGGGAGGCGGAGGTTGCAGTGAGCCGAGATCGCGCC
                             210       220       230       240       250
HSU14574.seq Sx       GAATCGCTTGAACCCGGGAGGCGGAGGTTGCAGTGAGCCGAGATCGCGCC  240
HSU14573.seq Sq       GAATCGCTTGAACCCGGGAGGCGGAGGTTGCAGTGAGCCGAGATCGCGCC  242
HSU14572.seq Sp       GAATCGCTTGAACCCGGGAGGCGGAGGTTGCGGTGAGCCGAGATCGCGCC  242
HSU14571.seq Sc       GAATCGCTTGAACCCGGGAGGCGGAGGTTGCAGTGAGCCGAGATCGCGCC  240
HSU14567.seq J        GGATCGCTTGAGCCCGGGAGGTCGAGGCTGCAGTGAGCCGTGATCGCGCC  242

                      ACTGCACTCCAGCCTGGGCGACA-GAGCGAGACTCCGTCTCAAAAAAAA
                             260       270       280       290       300
HSU14574.seq Sx       ACTGCACTCCAGCCTGGGCGACA-GAGCGAGACTCCGTCTCAAAAAAAA   290
HSU14573.seq Sq       ACTGCACTCCAGCCTGGGCAACAAGAGCGAAACTCCGTCTCAAAAAAAA   291
HSU14572.seq Sp       ATTGCACTCCAGCCTGGGCAACAAGAGCGAAACTCCGTCTCAAAAAAAA   291
HSU14571.seq Sc       ACTGCACTCCAGCCTGG-CGACA-GAGCGAGACTCCGTCTCAAAAAAAA   287
HSU14567.seq J        ACTGCACTCCAGCCTGGGCGACA-GAGCGAGACCCTGTCTCAAAAAAAA   290
```

Figure 1. Consensus sequence of known Alu repeat sequences, generated by performing an alignment analysis of five Alu consensus sequences representing the two family branches J and S

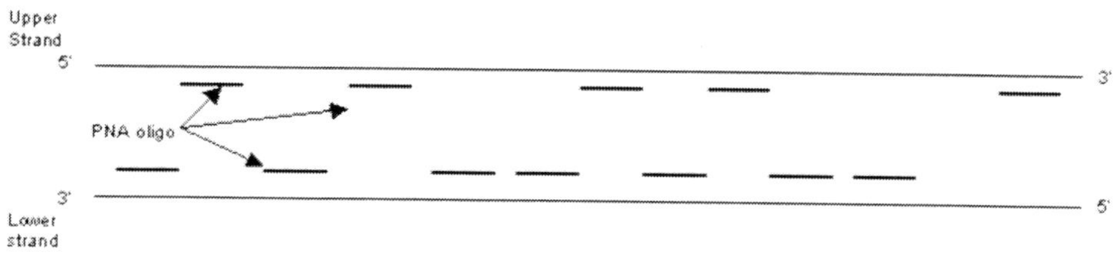

Figure 2. Positioning of probes on the individual strands within complementary strands of a unit repeat of an Alu repeat consensus sequence. The design of the PNAs is intended to cover as much as possible of one or the other of the complementary strands of the genomic nucleic acid. This approach can also serve to keep the strands separated

For this purpose, a fluorescein labelled PNA was synthesized for each candidate. The various PNA constructs were then evaluated in a PNA-FISH assay described below. The most important of the parameters to be analysed was the R-banding potential but the relative intensity of the R-banding signals was also examined. In each of the categories, the R-banding potential and signal intensity for each PNA probe was assigned a score from the minimum value of 0 to a maximum value of 6. To select for PNA probes that were most suitable for use in a PNA mixture for blocking of the Alu repeats, all PNAs having an R-banding potential value of less than 3 were eliminated. The procedure was used to screen a total of 55 unique PNA oligos, of which the 12 PNAs identified in Table 1 were preferred because of their superior ability to bind to genomic nucleic acid. The data in Table 1 is based on an average of 4-6 independent evaluations of each probe.

Table 1.

Seq No.	Position[a]	Nucleobase Sequence	R-band Score	Signal Intensity
1	9-25	GGCCGGGCGCGGTGGCT	4	5
2	26-43	GCTGGGATTACAGGCGTG	5	6
3	49-64	GGGAGGCCGAGGCGGG	5	5
4	82-102	GCCAGGCTGGTCTCGAACTCC	4	4
5	112-132	GAAACCCCGTCTCTACTAAAA	3	3
6	145-160	GCCGGGCGTGGTGGCG	5	4
7	163-180	TAGCTGGGATTACAGGCG	6	5
8	184-200	GGGAGGCTGAGGCAGGA	6	6
9	201-221	CCTCCCGGGTTCAAGCGATTC	3	3
10	228-244	TTGCAGTGAGCCGAGAT	3	3
11	253-273	TGCACTCCAGCCTGGGCGACA	4	4
12	292[c]	[b]TT(k)TTTTT(k)TTTLysOLysOTTT(k)TTTTT(k)TT	4	5

[a] Refers to the relative positions in the Alu consensus sequence depicted in Figure 1.
[b] Triplex maker construct. k = D-Lysine; Lys = L-lysine; O = 8-amino-3,6-dioxaoctanoic acid.
[c] Hybridises to the variable polyA tail of the alu sequence [22]

Representative examples of fluorescein labelled PNA oligos that obtained a high score in the R-banding evaluation approach (Seq. No. 2 and Seq. No. 8) and constructs that only just passed the limit value (Seq. No. 5 and Seq. No. 10) are shown in Figure 3. These images demonstrate that vast differences in performance exist for the various probes examined and therefore demonstrate that testing is a suitable way to determine which sequences were most effective at hybridising to the randomly distributed repeat sequences of genomic nucleic acid.

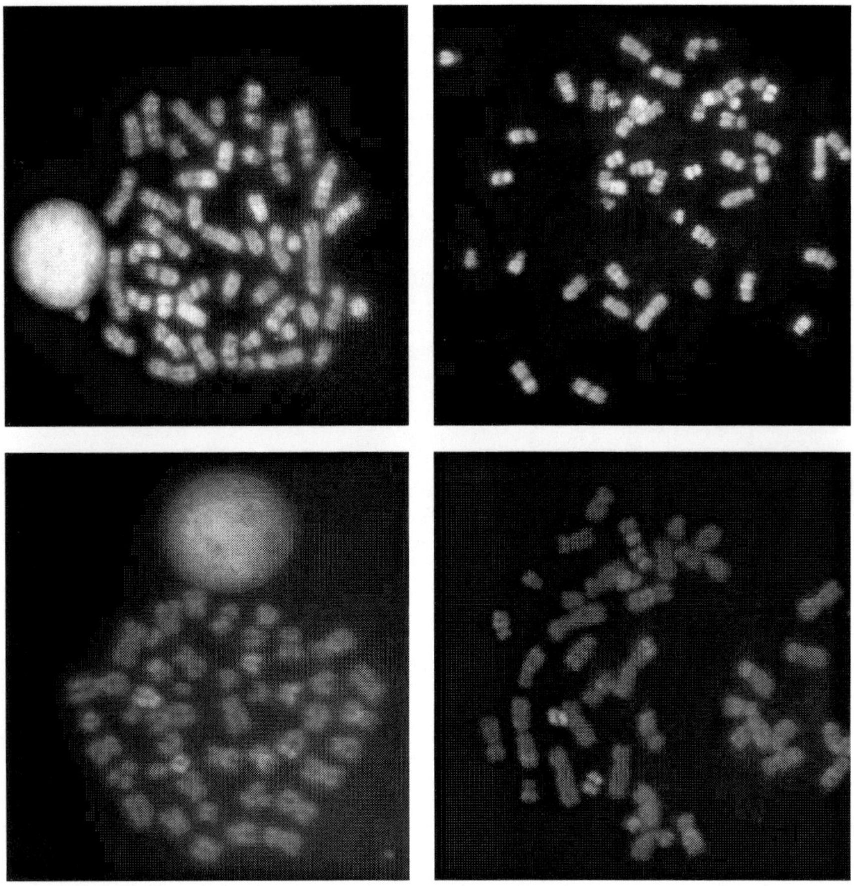

Figure 3. Digital images of interphase nuclei and metaphase spread of human chromosomes treated with labelled PNA oligos that either obtained a high R-banding score (upper panel) or that just passed the defined lower limit for a suitable R-banding pattern score (lower panel)

Effect of Sequence Variation in the Selection of Alu Repeat PNAs

The Alu elements are dominated by the J and S family branches [14,16]. Especially noteworthy is the Sx subfamily, whereas the Y family branch constitutes less than 0.5% of the total human Alu elements. In order to test the sensitivity by which the R-banding approach can detect the frequency of a specific sequence within the Alu elements, comparisons of almost identical PNA oligomer constructs were performed. The constructs to be compared were directed towards identical positions within the Alu consensus sequence (Figure 1) but the specific sequence design was based on either the J and S family consensus or the Y family consensus. The images presented in Figure 4 were obtained using PNA oligos that are complementary to positions 82-102 and 201-221 of the sequence shown in Figure 1.

Within positions 82-102 the J and S consensus differ from the Y consensus by a T → A transversion at position 86 and a G → T transversion at position 96. Similarly, within position 201-221 a single nucleotide polymorphism correspondingly consists of a T → G transversion at position 208. The sequence of the various PNA oligomer probes used in the experiments, as well as the respective R-band and Signal intensity scores of the examined oligomer constructs (Figure 4), are presented in Table 2. R-band and Signal intensity scores represent average values from 8-10 independent experiments.

Table 2.

Seq. Id. No.	Alu consensus	Oligo sequence	Relative pos. in Figure 2	R-band	Signal intensity
4	J and S family	GCCAGG**C**TGGTCTCGA**A**CTCC	82-102	4	5
4B	Y family	GCCAGG**A**TGGTCTCGA**T**CTCC	82-102	2	2
11	J and S family	CCTCCCGGGTTCA**A**GCGATTC	201-221	3	3
11B	Y family	CCTCCCGGGTTCA**C**GCGATTC	201-221	1	1

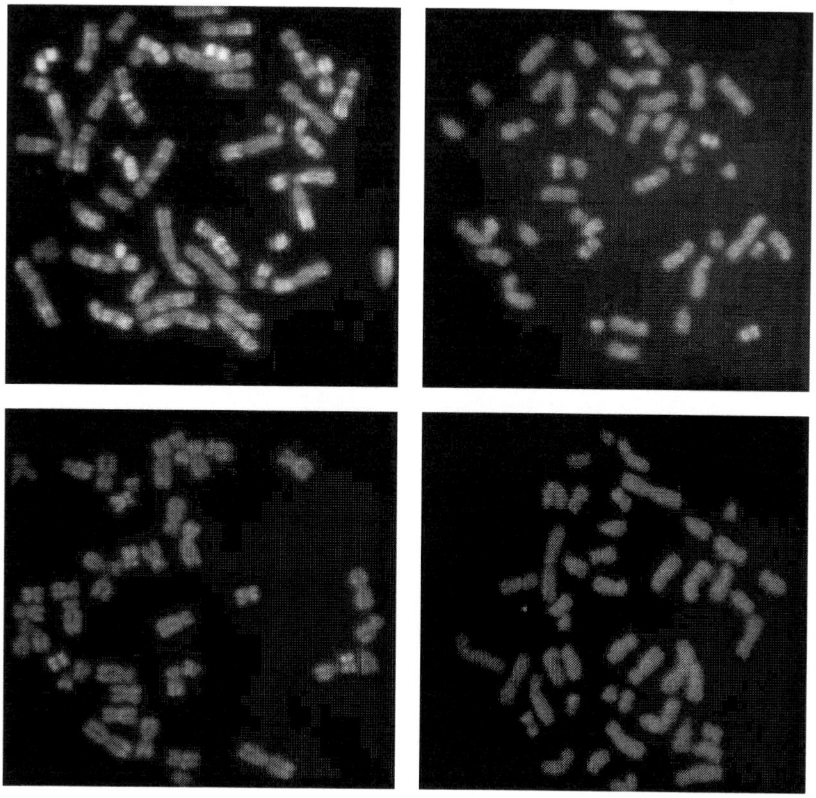

Figure 4. Digital images of a metaphase spread of human chromosomes treated with labelled PNA oligos designed from the J and S family consensus (upper panel) or the Y family consensus (lower panel). The Alu-banding is reduced when the Alu repeat PNAs are designed to target the more infrequent Y family branch

As illustrated in Table 2 and Figure 4, small changes (a single point mutation in one probe set and a double mutation in the other probe set) in the sequence of the PNA oligos significantly affects the efficiency by which the R-bands are formed. It is not known whether the R-band signals from hybridisation experiments involving the Seq. No. 4B and Seq. No. 11B constructs (Y family consensus) reflect a low frequency of the corresponding genomic sequences, or rather that they hybridise to the consensus sequences of the J and S family with a reduced affinity. Nevertheless, the data reinforce the argument that actual testing should be performed in order to determine which of all possible sequences will produce the most complete hybridisation to the randomly distributed repeat sequences and thereby presumably produce the most efficient blocking PNAs.

Suppression of Undesired Background Using a PNA Oligo Mix

The Alu-banding approach facilitated the identification of sequences that seem to be present in the human genome with a high frequency. If Alu repeats are a major reason for non-target hybridisation of large probes of genomic origin, a mixture of the identified PNA oligos should be able to suppress this undesired hybridisation background. Therefore, unlabelled versions of the PNA oligos listed in Table 1, were prepared, and mixed together. This "PNA Oligo Mix" was used in hybridisation experiments together with various DNA probes of genomic origin on either cytogenetic preparations or tissue sections.

The blocking approach was first tested in a one-colour FISH experiment. A green signal was generated by a fluorescein labelled *HER2* DNA probe, hybridised to metaphase spreads and interphase nuclei using the PNA Oligo Mix as blocking agent which only blocks the alu sequences (Figure 5, upper left). This result was compared with the standard blocking reagent, Cot-1 DNA (Human Cot-1 DNA from Gibco BRL, Life Technologies) which unselected blocks all repetitive sequences including centromere sequences (Figure 5, upper middle). In a separate sample, neither the PNA Oligo Mix nor the Cot-1 DNA was added (Figure 5, upper right). For the *HER2* experiments, COS clones covering 100 kb of a chromosome 17 region which include the *HER2* gene were used.

The approach was also tested in a dual-colour FISH experiment. The green and red signals were generated by two genomic probes derived from PAC clones covering 90 kb on each side of *MLL* gene (11q23) major breakpoint region (mbr) that were hybridised to metaphase spreads and interphase nuclei. One clone was labelled with fluorescein, the other with Cy3. A representative microscope image of the resulting sample can be seen in Figure 5 (middle left). This result was compared with Cot-1 DNA (Figure 5, middle). In a separate sample, neither the PNA Oligo Mix nor the Cot-1 DNA was added (Figure 5, middle right).

In another dual-colour FISH experiment, the red signal was generated by a genomic DNA probe while the green signal was derived from a mixture of labelled PNA probes. The labelled DNA probe is composed of COS clones covering approximately 100 kb of a region which include the *TOP2A* gene (17q21-22) labelled with Texas Red. The fluorescein labelled PNA probe was a mixture of 7 PNAs specific for the chromosome 17 α-satellite sequences (CEN-17 PNA probe). The probes were hybridised to metaphase spreads and interphase nuclei. Figure 5 (lower left) demonstrates the ability for the PNA Oligo Mix to block unspecific background when detecting the *TOP2A* gene on normal metaphase spreads using the green CEN-17 PNA probe as a reference. Green and red signals are clearly visible using the PNA mixture as a blocking reagent (Figure 5, lower left), whereas the green signals are somewhat dimmer when using Cot-1 DNA as a blocking reagent (Figure 5, lower middle). Specific signals cannot be distinguished from background in the absence of any blocking reagent (Figure 5, lower right).

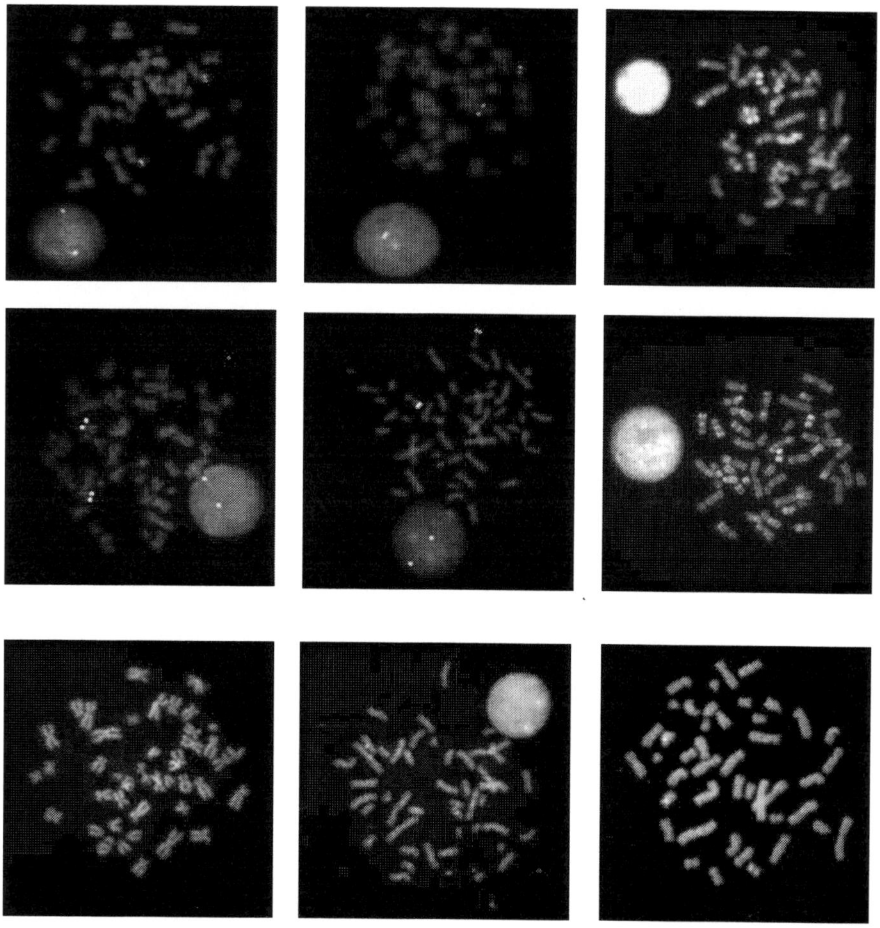

Figure 5. Digital images of interphase nuclei and metaphase spread of human chromosomes hybridised with single-colour *HER2* probe (upper panel), dual-colour *MLL* probes (middle panel), and *TOP2A*/CEN-17 probes (lower panel) combined with either: 1) a blocking mixture of unlabelled PNA oligos (left); 2) Cot-1 DNA blocker (middle) or is otherwise not treated with a blocking reagent (right)

With reference to Figure 5, it is apparent that in all three cases the mixture of PNA oligos appears to work as well as, if not better than, Cot-1 DNA. It is also noteworthy that the absence of any blocking agent (e.g. PNA Oligo Mix or the Cot-1 DNA) results in the formation of a hybridisation background with an R-banding pattern, thereby indicating that the undesired hybridisation is mainly caused by the presence of Alu repeats. The intensity of this background also seriously hinders the visualization of the single locus specific signals such that it must be removed in order to facilitate the performance of an accurate and reproducible assay.

Suppression of Undesired Background in Tissue Sections Using PNA Oligo Mix

The most frequent genetic changes observed in solid tumours (e.g. carcinomas) are amplifications and deletions of genes or whole chromosomal regions (amplicons). The FISH method can be used to detect these gene copy number changes in tissue sections of paraffin embedded tumours, which are the archival material that are most often available from solid tumours.

The PNA blocking mixture was used to suppress the undesired background staining from different probes, targeting genes known to be amplified or deleted in cancer cells. The experiments were performed on breast cancer sections.

In the first experiment (Figure 6, upper panel), the DNA probes for the *HER2* gene were COS clones labelled with Texas Red and mixed with the PNA Oligo Mix and the CEN-17 PNA probe described earlier. In non-amplified tissue sections (Figure 6, upper left) an equal number of green and red signals are present, while amplified tissue sections (Figure 6, upper right) have more red than green signal.

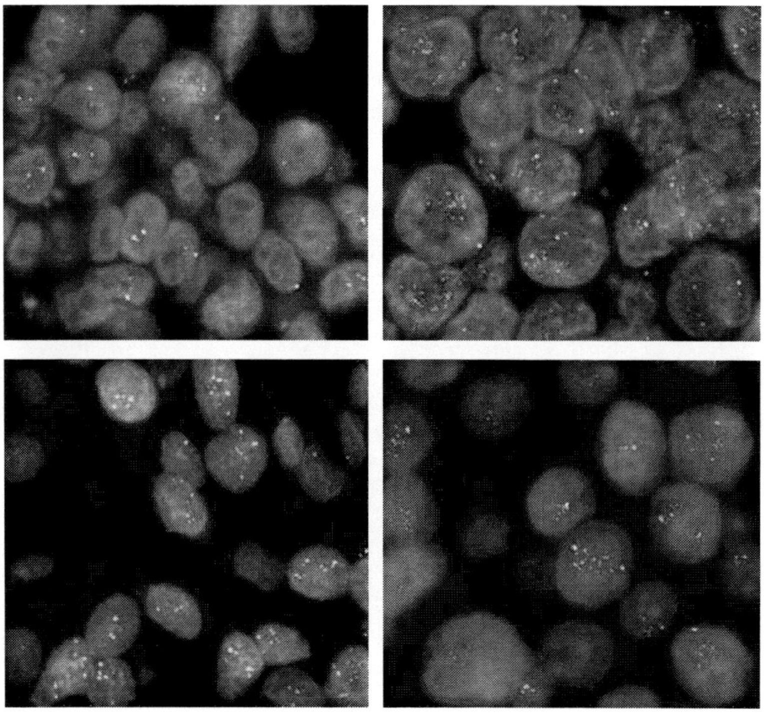

Figure 6. Digital images of *HER2*/CEN-17 probes on paraffin embedded tissue sections of a breast carcinoma with normal (upper left) and *HER2* amplification (upper right), respectively; *TOP2A*/cen-17 on paraffin embedded tissue sections of a breast carcinoma with *TOP2A* deletion (lower left) or *TOP2A* amplification (lower right), respectively

Amplification and deletion of the topoisomerase IIα gene (*TOP2A*) has been reported in breast cancer [33]. The PNA Oligo Mix was used to suppress the undesired background staining from a *TOP2A* DNA probe and a CEN-17 PNA probe. The *TOP2A* DNA probe, labelled with Texas Red and the CEN-17 PNA probe labelled with fluorescein were tested on formalin fixed, paraffin embedded tissue from a mammary carcinoma with *TOP2A* deletion (Figure 6, lower left) and a mammary carcinoma with amplification (Figure 6, lower right). The *TOP2A* DNA probe and CEN-17 PNA probe were mixed with the PNA Oligo Mix.

PNA Oligos as Centromere Directed Reference Probes

Design of FISH probes for genetic characterization of solid tumours may include a reference probe in addition to the primary gene-targeting probe. Use of a reference probe introduces a ratio between the two probes, rather than a gene signal number alone, as analysis output. The ratio information is beneficial as it enables discrimination between genetic events like amplification and polysomies or deletions and loss of genomic material during sample preparation.

A FISH reference probe is traditionally located on the chromosome harbouring the gene target of interest, and, whenever possible, the reference probe is located at the non-coding centromere region. The centromeres of higher eukaryotes are characterized by the massive presence of tandemly repeated DNA. The magnitude of repeated centromere DNA varies on each chromosome from hundreds of kilobases to tens of megabases. The central domain of the centromeres harbours the α-satellite family, which make up the most abundant repeat DNA found at the centromeres of all human chromosomes [34-37].

α-satellite is composed of 171bp monomers tandemly arranged in a head to tail fashion [38,39]. These monomer arrays spans millions of base pairs and constitute 1-5% of the individual chromosomes. The monomers are organized into higher order structures (Figure 7) and based on a complex relationship of both sequence and structure (reviewed by [40,41]) α-satellite DNA has been divided into subfamilies that are largely chromosome specific (reviewed by [42]). Generally, the DNA sequences of the individual monomers within a higher order structure (Figure 7 B) diverge in the order of 20-40% and the sequence divergence between α-satellite subfamilies is also significant (> 15-30%) [42].

The high copy number of individual α-satellite monomer types on the different centromeres implies that even a single oligo sequence may result in hybridisation to thousands of targets within a well-defined area. The premise is that the centromere specificity is defined not only at the higher order level (Figure 7 B) but also at the monomer level alone. Further, successful use of an approach based on monomer specificity is dependent on the discriminating capacity of the oligo. Complementary PNA/DNA strands bind more strongly than corresponding DNA/DNA duplexes and, perhaps more importantly, mismatch in a PNA/DNA duplex is more destabilizing than a mismatch in a DNA/DNA duplex [43]. The high ΔT_m of PNA/DNA duplexes therefore points at PNA oligos as an attractive means for discrimination between chromosome specific α-satellite monomers. Single base discrimination between the α-satellite repeats CENP-B has been demonstrated with PNA FISH [44] and unique chromosome identification by PNA FISH has been reported for several chromosomes [45]. These studies demonstrate that chromosome specific variation at the monomer sequence level alone may be sufficient for functional discrimination between chromosomes when PNA is used as a probe.

Like Chen and co-workers [45] we have used an approach that includes computing of genomic sequences from databases to search for chromosome specific α-satellite monomer sequences. By examination of numerous 15-20bp PNA oligo design candidates we have identified specific PNA oligos towards the α-satellite repeats of several chromosomes. Further, in order to increase signal intensity of the individual oligos, we have labelled each PNA oligo with four fluorochromes (mostly 5-Carboxy fluorescein). The resulting PNA oligo based fluorescent probes have proven highly efficient as illustrated in Figure 8.

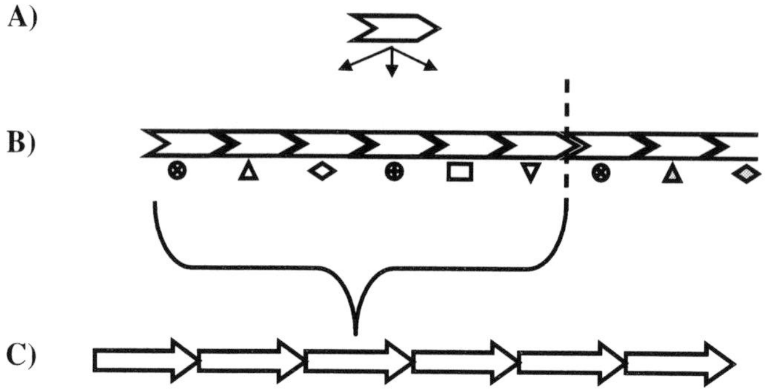

Figure 7. Fundamental organisation of α-satellite repeats: A) Basic 171bp monomer B) Example of chromosome specific variation and organisation of basic monomers into higher-order structures C) Long-range repetition of higher-order structure. Principle from Choo K.H.A. 1997

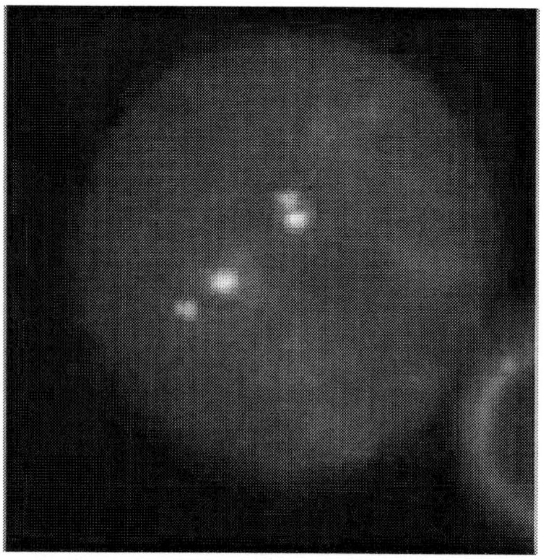

Figure 8. FISH analysis on a interphase nucleus illustrating the relative signal intensity obtained from a single PNA oligo towards a chromosome 7 α-satellite monomer (green) as compared to the signal obtained from a genomic probe towards the EGFR gene (red) representing about 4000 Texas red fluorochromes

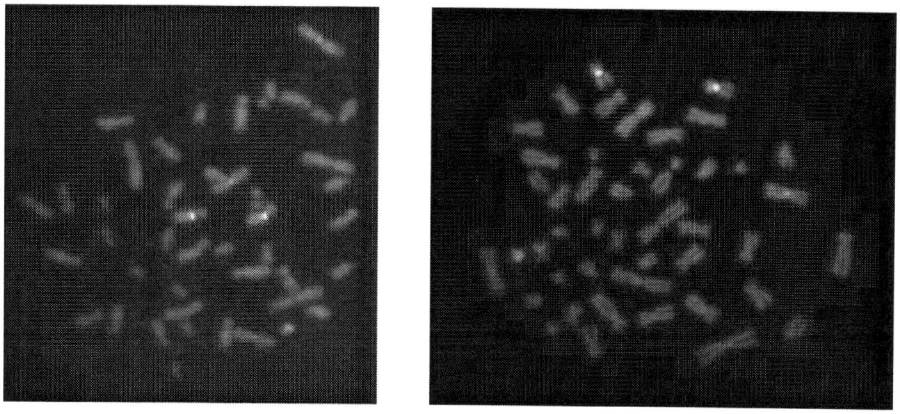

Figure 9. FISH on metaphase spreads demonstrating the specificity of high signal intensity fluorescein labelled PNA oligos towards α-satellite monomers. The examples show single PNA oligos towards monomers located on chromosome 7 (left) and chromosome 11 (right)

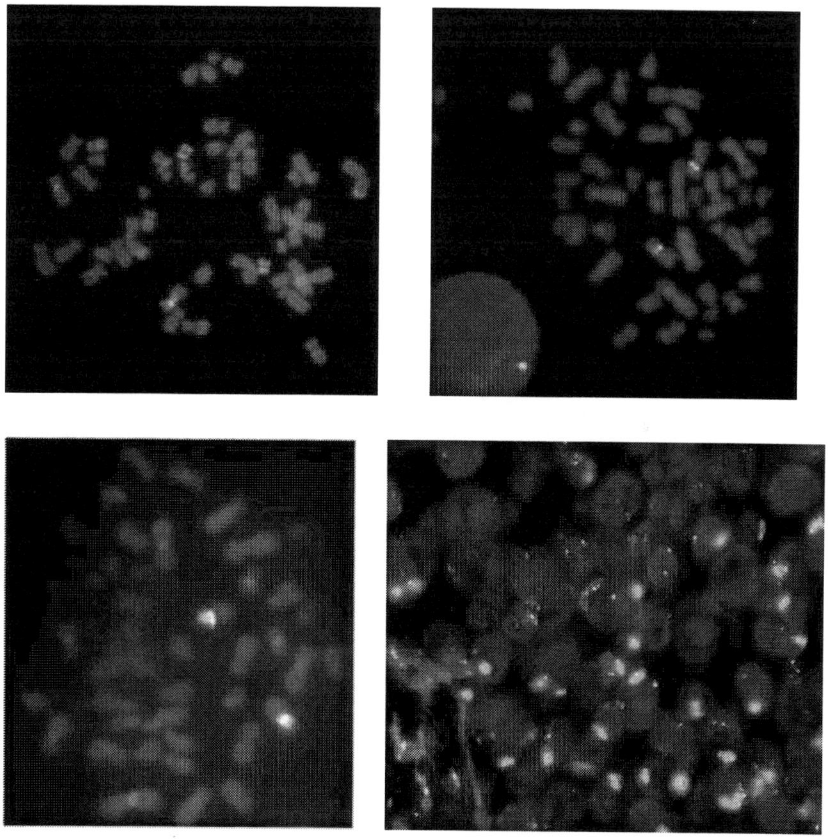

Figure 10. Texas red labelling of DNA from a BAC clone covering the *EGFR* gene hybridised to a metaphase spread (upper left) without PNA oligo mixture for blocking of unspecific hybridisation (upper right) including PNA oligo blocking mixture. *EGFR*/CEN-7 DNA/PNA FISH probe mixture applied on a metaphase spread sample (lower left) and FFPE tissue from a NSCLC patient (lower right) – EGFR genomic probe is Texas red labelled, the PNA oligo based centromere 7 reference probe is fluorescein labelled (green)

As apparent from Figure 8, the signal intensity from a single, fluorescein labelled, PNA oligo towards a chromosome 7 α-satellite monomer sequence, matches the signal intensity from a Texas Red labelled genomic probe representing approximately 4000 fluorescence molecules. The high signal intensity does not compromise the signal specificity as illustrated on the metaphase hybridisation examples in Figure 9. In a final probe mixture, e.g. *EGFR*/CEN-7, the PNA based reference probe is composed of a mixture of several labelled PNA oligos and the concentration of this mixture is adapted to produce a signal intensity which is balanced with the gene targeted genomic probe signal (Figure 10, lower right).

A DNA/PNA Based FISH Probe Mixture for Detection of *EGFR* Gene Copy Numbers

The gene encoding epidermal growth factor receptor, *EGFR*, is located close to the chromosome 7 centromere on position 7p11.2. The genomic sequence of *EGFR* spans 188kb. Computer analysis for repeated elements within the *EGFR* genomic sequence reveals a relatively low presence of repeats, in total around 15%, of which the major part is made up by SINEs and LINEs. Despite the low frequency of repeated elements, labelling of a BAC clone covering the *EGFR* genomic sequence results in significant unspecific staining as illustrated on the metaphase spread hybridisation in Figure 10 (upper left). Figure 10, upper right, shows that blocking of this unspecific hybridisation is efficiently achieved by use of the above described unlabelled PNA blocking mixture.

Several recent publications indicate a clinical value of *EGFR* FISH for prediction of response to EGFR inhibitors in NSCLC and colorectal cancers [46-48]. In the context of NSCLC, it appears that amplifications as well as polysomies, in various combinations of intensity and spread, make up relevant scoring parameters for the identification of EGFR inhibitor responders (the Cappuzzo scoring key) [46,47,49]. The Cappuzzo key stresses the importance of a well-functioning reference probe. The metaphase spread FISH in Figure 10 (lower left) includes a PNA based reference probe (green) in addition the *EGFR* genomic probe (red) and illustrates the close location of the *EGFR* gene to the chromosome 7 centromere. The centromere 7 PNA probe is generated by use of the novel 4 fluorescein per PNA oligo technology and contains a mixture of centromere 7 specific oligos. The high specificity and signal-intensity properties of the synthetic PNA oligos permit a precise titration of the reference signal and thereby allows control of the final balance between the gene-targeted probe and the reference probe signals. An example of how the final DNA/PNA FISH probe mixture (*EGFR*/CEN-7, Y5500) for detection of *EGFR* gene copy no. works on FFPE tissue is illustrated in Figure 10 (lower right). The tissue example shows a NSCLC tissue section dominated by heavy clustered amplification of the *EGFR* gene (red splotch) but also cells with normal *EGFR* gene status (upper left part of the picture). The normal cells illustrate the balance between the red gene signal and the green reference signal, whereas the red signal in cells with clustered amplification is so massive that it covers the green reference signal and thereby converts it to a yellow spot. All hybridisations on metaphase spreads and on tissue sections in Figure 10 were performed by use of the cytology and histology protocols described above in the Methods section.

Use of PNA to Suppress the Undesired Background Staining when Detecting Genetic Translocations

Specific chromosomal rearrangements, in particular translocation (exchange of a piece from one chromosome onto another chromosome) are common occurrences in cancer, and are observed in the great majority of human haematopoietic malignancies. Such characteristic chromosomal abnormalities can be of diagnostical, prognostical or classification relevance in leukaemia and lymphomas assayed on cytology samples, cytogenetic preparations or tissue sections. One method of detecting such genetic translocation is split signal FISH. In split signal FISH, probes are labelled with fluorescent signals that are detectable, generally by microscopic viewing of colours. Two DNA sequences that flank a specific breakpoint region of interest are labelled with different fluorescent dyes. The combined probes are homologous to the breakpoint flanking target DNA sequences of the specific locus of interest. By forming hybrids between the locus sequences and the FISH probe a distinct hybridisation pattern is observed in gene translocation negative and positive cells. The PNA blocking mixture has been used to suppress the unspecific staining from a split signal FISH *CCND1* probe when used on normal metaphase spreads, interphase nuclei, and FFPE tissue. The upstream part of the *CCND1* probe was labelled with FITC and the downstream part of the *CCND1* probe with Texas Red, mixed with the PNA oligo mixture (Figure 11 upper left).

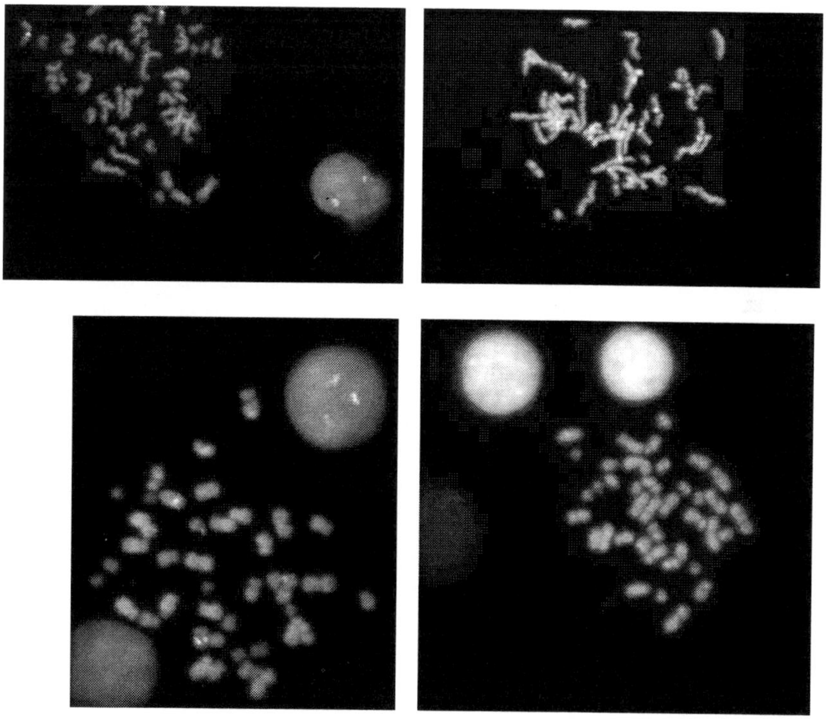

Figure 11. Digital images of interphase nuclei and metaphase spreads hybridised with CCND1 translocation probe on normal lymphocytes (upper panel) and a cell line (lower panel) harbouring a translocation. Experiments were performed with (left panel) and without (right panel) the PNA oligo mix

For comparison Figure 11, upper right, shows the same experiment without the PNA oligo mix added. The CCND1 gene is located on chromosome band 11q23. Figure 11 (upper left) shows normal CCND1 configurations on metaphases with a yellow signal located on both chromosome 11. Two yellow signals in interphase nuclei indicate two normal CCND1 loci.

The CCND1 probes were also hybridised to metaphase spreads of the cell line Granta 519 [50] that harbours a translocation t(11;14)(q23,q32). The CCND1 probes were labelled as above and mixed with either the PNA oligo mixture (Figure 11, lower left). Figure 11, lower left, shows one green signal located der(14) and a red signal on der(11), indicating a split of the upstream CCND1 and downstream CCND1 probes, indicating a translocation. A yellow signal on the other allele of chromosome 11 indicates a normal CCND1 configuration. For comparison Figure 11, lower right, shows the same experiment without the PNA oligo mix.

PNA Probes in Sequential FISH Using the Concept of Competitive Displacement

Recently the PNA probes have been introduced in single cell FISH in connection with Preimplantation Genetic Diagnosis (PGD) on blastomere nuclei from human embryos. Preimplantation Genetic Diagnosis of chromosomes has so far primarily been based on FISH with DNA probes. The method has, however, the drawback that only a limited number of chromosomes can be analysed in one FISH procedure [51]. To compensate for this it has been confirmed that sequential FISH can be applied to blastomeres performing several rounds of FISH with different DNA probes [52-54]. A dilemma when performing repeated FISH cycles using DNA probes is the high temperatures needed to denature the target- as well as probe DNA before binding and especially for re-annealing the probes after each FISH cycle. The high temperature compromises the integrity of the target DNA and thus increases the risk of false results in subsequent cycles [30].

The lengths of PNAs used in hybridisation studies are relatively shorter than DNA or RNA probes and due to the chemically modified peptide backbone the PNA probes are competent to bind to target DNA at relative low temperatures. In 2004, Paulasova *et al* [55] demonstrated that blastomere nuclei are well suited for FISH using PNA probes in at least two cycles, but the experiment was performed with denaturation temperatures of 73 °C. In 2005, Agerholm *et al* [56] demonstrated that PNA probes could bind and generate strong clear FISH signals in human blastomere nuclei at 55 °C. More importantly Agerholm *et al* [56] also demonstrated that due to the synthetic nature of PNA probes it is no problem to synthesize unlabelled oligos of exactly the same composition as the labelled probe. This gives the possibility of introducing the concept of competitive displacement. In competitive displacement, unlabelled probe with the same composition as already bound and labelled probe will compete for annealing at the same sequence of the target DNA. Excess unlabelled probe will therefore destabilize the already bound and labelled probe resulting in a reduction of the denaturation temperature necessary for dissociation of already bound probe before the next FISH cycle in the sequential FISH set-up.

The temperature needed for binding and displacement of PNA probes in the presence of unlabelled probe was demonstrated on lymphocyte metaphases where the signal left over was measured in the presence or absence of unlabelled probe in the following FISH cycle. PNA probes mixtures specific for chromosome 1, 16, 17, 18, X and Y were obtained from Applied Biosystems together with a PNA probe mixture with specificity towards both chromosome 13 and chromosome 21. Each chromosome probe mixture is composed of one to several oligomers with sequence complementarity to chromosome specific sequences of the relevant alpha monomer. The probe mixtures were available both as labelled and as unlabelled mixtures. The concept of competitive displacement means that the second and third probe set used in the study is composed of a combination of labelled probes for the chromosomes analysed in the cycle and unlabelled probes for the chromosomes in the previous cycle. As a consequence the different probe sets must be applied in a fixed order when performing sequential FISH.

Probe set A has labelled probes for chromosome 13 and 21 and no unlabelled probe. Probe set B has labelled probes for chromosome 1, 16, 17 and unlabelled probes for chromosome 13 and 21. Probe set C has labelled probes for chromosome 18, X and Y and unlabelled probes for chromosome 1, 16 and 17. The fluorochrome labelling, number of oligomers and concentration of the various probes and the composition of the probe mixtures are described in table 1 of Agerholm et al [56].

For all probe sets, presence of unlabelled probe in the denaturation mixture increases the dissociation of bound, labelled probe at a range of temperatures. A typical example using the 13/21 probe set is shown in Figure 12. Although the individual curves varied somewhat, FISH signals were removed from all chromosomes at a denaturation temperature of 50° C, when unlabelled probe is present. In a separate series it was ensured that denaturation at 55° C with unlabelled probe did not have a negative effect on the subsequent FISH signals (data not shown).

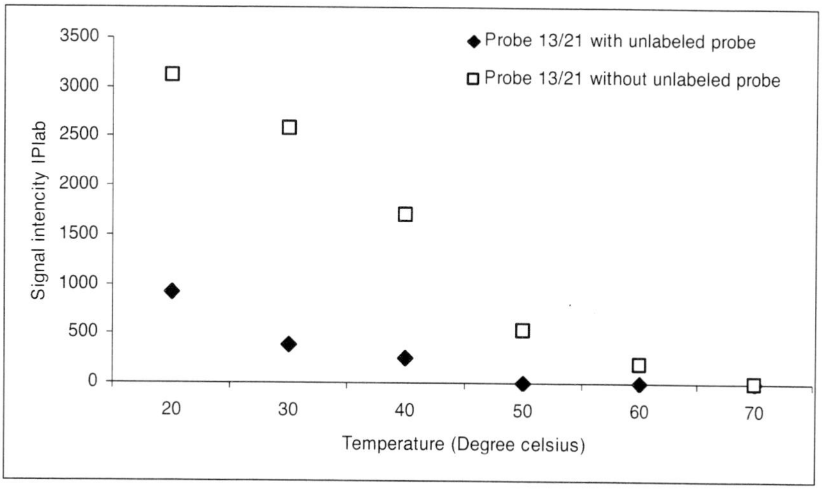

Figure 12. Correlation between denaturation temperature and signal intensity in melting experiments with and without presence of unlabelled probe

The experiments demonstrated that unlabelled PNA probe does destabilize the base pairing and results in easy displacement of bound PNA probe from a previous cycle. The destabilization depends equally on the temperature and the presence of unlabelled probe with the same composition as the bound and labelled (Figure 12). In the presence of unlabelled probe the melting temperature was, for all chromosomes, around 50ºC while without unlabelled probe in the following cycle the destabilization was already compromised at 60ºC. In a separate series, it was established that the displacement was not increased by higher concentrations of unlabelled probe in the displacement. Thus an increase in concentration of unlabelled probe from 5 nM to 100 nM did not alter the temperature profiles of displacement.

The fact that a lower temperature is sufficient to remove bound PNA probes is particularly important in single cell FISH. Repeated FISH cycles are necessary for the enumeration of several chromosomes, and such a procedure will always require a denaturation step in each cycle. The presence of unlabelled probe in the following FISH cycle favours denaturation at lower temperature, thereby minimizing FISH errors in up to five PNA FISH cycles.

Another advantage when using PNA probes for enumeration of human blastomeres is the timeframe. Due to the higher affinity of the DNA-PNA duplex the hybridisation can be performed at room temperature over a short period e.g. 30 minutes. As a consequence of the fast hybridisation kinetic of the PNA probes 5 cycles of FISH-PNA will require only 5-6 hours and the pre-embryos can be transferred at the same day as the biopsy is performed. This is important in connection with PGD where the results from the preimplantation genetic analyses should be available as fast as possible.

Conclusion

The above mentioned examples demonstrate some of the characteristics of PNA oligos that makes them superior to DNA oligos in specific applications. The PNA is superior in applications targeting repetitive sequences in the human genome and can be used both in labelled and unlabelled versions. PNA oligos are especially superior to DNA oligos in binding, displacement and blocking reactions.

References

[1] Kallioniemi,A., Kallioniemi,O.P., Sudar,D., Rutovitz,D., Gray,J.W., Waldman,F. and Pinkel,D. (1992). Comparative Genomic Hybridization for Molecular Cytogenetic Analysis of Solid Tumors. *Science*, *258*, 818-821.

[2] Eils,R., Uhrig,S., Saracoglu,K., Satzler,K., Bolzer,A., Petersen,I., Chassery,J.M., Ganser,M. and Speicher,M.R. (1998). An optimized, fully automated system for fast and accurate identification of chromosomal rearrangements by multiplex-FISH (M-FISH). *Cytogenetics and Cell Genetics*, *82*, 160-171.

[3] Macville,M., Veldman,T., PadillaNash,H., Wangsa,D., OBrien,P., Schrock,E. and Ried,T. (1997). Spectral karyotyping, a 24-colour FISH technique for the identification of chromosomal rearrangements. *Histochemistry and Cell Biology*, *108*, 299-305.

[4] Tkachuk,D.C., Westbrook,C.A., Andreeff,M., Donlon,T.A., Cleary,M.L., Suryanarayan,K., Homge,M., Redner,A., Gray,J. and Pinkel,D. (1990). Detection of Bcr-Abl Fusion in Chronic Myelogeneous Leukemia by Insitu Hybridization. *Science*, *250*, 559-562.

[5] Britten,R.J. and Kohne,D.E. (1968). Repeated Sequences in Dna. *Science*, *161*, 529

[6] Novick,G.E., Batzer,M.A., Deininger,P.L. and Herrera,R.J. (1996). The mobile genetic element Alu in the human genome. *Bioscience*, *46*, 32-41.

[7] Brosius,J. (1991). Retroposons - Seeds of Evolution. *Science*, *251*, 753

[8] Singer,M.F. (1982). Sines and Lines - Highly Repeated Short and Long Interspersed Sequences in Mammalian Genomes. *Cell*, *28*, 433-434.

[9] Korenberg,J.R. and Rykowski,M.C. (1988). Human Genome Organization - Alu, Lines, and the Molecular-Structure of Metaphase Chromosome Bands. *Cell*, *53*, 391-400.

[10] Rogers,J.H. (1985). The Origin and Evolution of Retroposons. *International Review of Cytology-A Survey of Cell Biology*, *93*, 187-279.

[11] Ausubel,F.M., Brent,R., Kingston,R.E., Moore,D.D., Seidman,J.G., Smith,J.A., and Struhl,K. (1996): *Current Protocols in Molecular Biology*. John Wiley and Sons, Canada.

[12] Hattori,M., Fujiyama,A., Taylor,T.D., Watanabe,H., Yada,T., Park,H.S., Toyoda,A., Ishii,K., Totoki,Y., Choi,D.K., et al (2000). The DNA sequence of human chromosome 21. *Nature*, *407*, 110.

[13] Dunham,I., Shimizu,N., Roe,B.A., Chissoe,S., Dunham,I., Hunt,A.R., Collins,J.E., Bruskiewich,R., Beare,D.M., Clamp,M., et al (1999). The DNA sequence of human chromosome 22. *Nature*, *402*, 489-495.

[14] Batzer,M.A., Deininger,P.L., HellmannBlumberg,U., Jurka,J., Labuda,D., Rubin,C.M., Schmid,C.W., Zietkiewicz,E. and Zuckerkandl,E. (1996). Standardized nomenclature for Alu repeats. *Journal of Molecular Evolution*, *42*, 3-6.

[15] Sherry,S.T., Harpending,H.C., Batzer,M.A. and Stoneking,M. (1997). Alu evolution in human populations: Using the coalescent to estimate effective population size. *Genetics*, *147*, 1977-1982.

[16] Britten,R.J. (1994). Evidence That Most Human Alu Sequences Were Inserted in A Process That Ceased About 30 Million Years Ago. *Proceedings of the National Academy of Sciences of the United States of America*, *91*, 6148-6150.

[17] Lansdorp,P.M., Verwoerd,N.P., vandeRijke,F.M., Dragowska,V., Little,M.T., Dirks,R.W., Raap,A.L. and Tanke,H.J. (1996). Heterogeneity in telomere length of human chromosomes. *Human Molecular Genetics*, *5*, 685-691.

[18] Taneja,K.L., Chavez,E.A., Coull,J. and Lansdorp,P.M. (2001). Multicolor fluorescence in situ hybridization with peptide nucleic acid probes for enumeration of specific chromosomes in human cells. *Genes Chromosomes and Cancer*, *30*, 57-63.

[19] Thisted,M., Just,T., Pluzek,K.-J., Hyldig-Nilsen,J.J. and Godtfredsen,S.E. (1996). Detection of immunoglobulin kappa light chain mRNA in paraffin sections by in situ hybridization using peptide nucleic acid probes. *Cell Vision*, *3*, 358-363.

[20] Just,T., Burgwald,H. and Broe,M.K. (1998). Flow cytometric detection of EBV (EBER snRNA) using peptide nucleic acid probes. *Journal of Virological Methods*, *73*, 163-174.

[21] Taneja,K.L. (1998). Localization of trinucleotide repeat sequences in myotonic dystrophy cells using a single fluorochrome-labeled PNA probe. *Biotechniques*, *24*, 472-476.

[22] Ullu,E. (1982). The Human Alu Family of Repeated Dna-Sequences. *Trends in Biochemical Sciences*, *7*, 216-219.

[23] Landegent,J.E., Dewal,N.J.I., Vanommen,G.J.B., Baas,F., Devijlder,J.J.M., Vanduijn,P. and Vanderploeg,M. (1985). Chromosomal Localization of A Unique Gene by Non-Autoradiographic Insitu Hybridization. *Nature*, *317*, 175-177.

[24] Landegent,J.E., Dewal,N.J.I., Dirks,R.W., Baas,F. and Vanderploeg,M. (1987). Use of Whole Cosmid Cloned Genomic Sequences for Chromosomal Localization by Non-Radioactive Insitu Hybridization. *Human Genetics*, *77*, 366-370.

[25] Britten,R.J., Graham,D.E. and Neufield,B.R. (1986). Analysis of repeating DNA sequences by reassociation. *Methods in Enzymology*, *29*, 363-413.

[26] Nielsen, K. V. Hyldig-Nielsen, J.J. and Williams, B.F. Methods, kits and compositions pertaining to the suppression of detectable probe binding to randomly distributed repeat sequences in genomic nucleic acid. (US 20030129626). 2002.

[27] Nielsen,K.V., Müller,S., Poulsen,T.S., Gabs,S., and Schønau,A. (2003): Combined use of PNA and DNA for fluorescence in situ hybridization (FISH). In: *Peptide Nucleic Acids, Protocols and Applications*, edited by P.E.Nielsen, pp. 227-260. Horizon Scientific Press, Norwich.

[28] Sheehan,D.C. and Hrapchak,B.B. (1980): *Theory and practice of histotechnology*. Mosby Company, St. Louis.

[29] Kiernan,J.A. (1981): *Histological and histochemical methods: theory and practice*. Pergamon Press, New York

[30] Wilton,L. (2005). Preimplantation genetic diagnosis and chromosome analysis of blastomeres using comparative genomic hybridization. *Human Reproduction Update*, *11*, 33-41.

[31] Claverie,J.M. and Makalowski,W. (1994). Alu Alert. *Nature*, *371*, 752

[32] Baldini,A. and Ward,D.C. (1991). Insitu Hybridization Banding of Human-Chromosomes with Alu-Pcr Products - A Simultaneous Karyotype for Gene-Mapping Studies. *Genomics*, *9*, 770-774.

[33] Olsen,K.E, Knudsen,H., Rasmussen,B.B, Balslev,E., Knoop,A., Ejlertsen,B., Nielsen,K.V., Schonau, A., Overgaard, J. (2004). Amplification of HER2 and TOP2A and deletion of TOP2A genes in breast cancer investigated by new FISH probes. *Acta Oncol, 43,* 35-42.

[34] Manuelidis,L. (1978). Complex and Simple Sequences in Human Repeated Dnas. *Chromosoma*, *66*, 1-21.

[35] Manuelidis,L. (1978). Chromosomal Localization of Complex and Simple Repeated Human Dnas. *Chromosoma*, *66*, 23-32.

[36] Mitchell,A.R., Gosden,J.R. and Miller,D.A. (1985). A Cloned Sequence, P82H, of the Alphoid Repeated Dna Family Found at the Centromeres of All Human-Chromosomes. *Chromosoma, 92*, 369-377.

[37] Willard,H.F. (1985). Chromosome-Specific Organization of Human Alpha Satellite Dna. *American Journal of Human Genetics, 37*, 524-532.

[38] Rosenberg,H., Singer,M. and Rosenberg,M. (1978). Highly Reiterated Sequences of Simiansimiansimiansimiansimian. *Science, 200*, 394-402.

[39] Wu,J.C. and Manuelidis,L. (1980). Sequence Definition and Organization of A Human Repeated Dna. *Journal of Molecular Biology, 142*, 363-386.

[40] Willard,H.F. and Waye,J.S. (1987). Hierarchical Order in Chromosome-Specific Human Alpha-Satellite Dna. *Trends in Genetics, 3*, 192-198.

[41] Willard,H.F. (1991). Evolution of alpha satellite. *Current Opinion in Genetics and Development, 1*, 509-514.

[42] Choo,K.H.A. (1997): *The Centromere*. Oxford University Press.

[43] Egholm,M., Buchardt,O., Christensen,L., Behrens,C., Freier,S.M., Driver,D.A., Berg,R.H., Kim,S.K., Norden,B. and Nielsen,P.E. (1993). Pna Hybridizes to Complementary Oligonucleotides Obeying the Watson-Crick Hydrogen-Bonding Rules. *Nature, 365*, 566-568.

[44] Chen,C., Hong,Y.K., Ontiveros,S.D., Egholm,M. and Strauss,W.M. (1999). Single base discrimination of CENP-B repeats on mouse and human Chromosomes with PNA-FISH. *Mammalian Genome, 10*, 13-18.

[45] Chen,C.F., Wu,B.L., Wei,T., Egholm,M. and Strauss,W.M. (2000). Unique chromosome identification and sequence-specific structural analysis with short PNA oligomers. *Mammalian Genome, 11*, 384-391.

[46] Tsao,M.S., Sakurada,A., Cutz,J.C., Zhu,C.Q., Kamel-Reid,S., Squire,J., Lorimer,I., Zhang,T., Liu,N., Daneshmand,M., et al (2005). Erlotinib in lung cancer - Molecular and clinical predictors of outcome. *New England Journal of Medicine, 353*, 133-144.

[47] Hirsch,F.R., Varella-Garcia,M., Bunn,P.A., Franklin,W.A., Dziadziuszko,R., Thatcher,N., Chang,A., Parikh,P., Pereira,J.R., Ciuleanu,T., et al (2005). Molecular analysis of EGFR gene copy number, EGFR expression and AKT activation status in advanced non-small-cell lung cancer (aNSCLC) treated with gefitinib or placebo (ISEL trial). *Clinical Cancer Research, 11*, 9031s.

[48] Moroni,M., Veronese,S., Benvenuti,S., Marrapese,G., Sartore-Bianchi,A., Di Nicolantonio,F., Gambacorta,M., Siena,S. and Bardelli,A. (2005). Gene copy number for epidermal growth factor receptor (EGFR) and clinical response to antiEGFR treatment in colorectal cancer: a cohort study. *Lancet Oncology, 6*, 279-286.

[49] Cappuzzo,F., Hirsch,F.R., Rossi,E., Bartolini,S., Ceresoli,G.L., Bemis,L., Haney,J., Witta,S., Danenberg,K., Domenichini,I., et al (2005). Epidermal growth factor receptor gene and protein and gefitinib sensitivity in non-small-cell lung cancer. *Journal of the National Cancer Institute, 97*, 643-655.

[50] Drexler,H.G. (2001): *The Leukemia-Lymphoma Cell Line Facts Book*. Braunschweig, Germany.

[51] Wilton,L. (2002). Preimplantation genetic diagnosis for aneuploidy screening in early human embryos: a review. *Prenatal Diagnosis, 22*, 512-518.

[52] Benadiva,C., Kligman,I. and Munne,S. (1996). Aneuploidy 16 in human embryos increases significantly with maternal age. *Fertility and Sterility*, *66*, 248-255.

[53] Gianaroli,L., Magli,M.C., Ferraretti,A.P. and Munne,S. (1999). Preimplantation diagnosis for aneuploidies in patients undergoing in vitro fertilization with a poor prognosis: identification of the categories for which it should be proposed. *Fertility and Sterility*, *72*, 837-844.

[54] Abdelhadi,I., Colls,P., Sandalinas,M., Escudero,T. and Munne,S. (2003). Preimplantation genetic diagnosis of numerical abnormalities for 13 chromosomes. *RBM online*, *6*, 226-231.

[55] Paulasova,P., Andreo,B., Diblik,J., Macek,M. and Pellestor,F. (2004). The peptide nucleic acids as probes for chromosomal analysis: application to human oocytes, polar bodies and preimplantation embryos. *Molecular Human Reproduction*, *10*, 467-472.

[56] Agerholm,I.E., Ziebe,S., Williams,B., Berg,C., Cruger,D.G., Petersen,G.B. and Kolvraa,S. (2005). Sequential FISH analysis using competitive displacement of labelled peptide nucleic acid probes for eight chromosomes in human blastomeres. *Human Reproduction*, *20*, 1072-1077..

In: PRINS and PNA Technologies in Chromosomal Investigations ISBN 1-60021-413-4
Editor: Franck Pellestor, pp. 169-175

Chapter 11

PNA Probes to Measure Repetitive DNA in Metaphase Chromosomes Using Fluorescence *In Situ* Hybridization

Peter M. Lansdorp*
Terry Fox Laboratory, British Columbia Cancer Agency, 675 West 10th Avenue, Vancouver, BC, V5Z 1L3, Canada
Department of Medicine, University of British Columbia, Vancouver, BC, V5Z 4E3, Canada

Abstract

Small (12-18 nucleotides) "peptide nucleic acid" (PNA) oligonucleotide probes labeled with suitable fluorescent dyes are excellent probes for fluorescence in situ hybridization (FISH) techniques. The main advantage of single stranded (ss) PNA over DNA or RNA oligonucleotide probes is that hybridization conditions can be used that will allow PNA but not RNA or DNA to anneal to selected target sequences. This is particularly important when FISH techniques target DNA repeats in metaphase chromosomes since complementary chromosomal DNA will rapidly displace short DNA or RNA probes from their specific target sequences. Such displacement will not occur when PNA probes are hybridized at low ionic strength conditions allowing *only* the PNA probe (with an uncharged "peptide" backbone) to anneal to target sequences defined by the base composition of the PNA probe. The result is a useful *quantitative* relationship between the amount of hybridized probe (measured by the amount of fluorescence) and the amount of repeat sequences at a specific site. Quantitative FISH (Q-FISH) using directly labeled $(CCCTAA)_3$ PNA specific for telomere repeat sequences has become the method of choice to measure the length of guanine-rich repeats at individual metaphase chromosome ends. Here the use of PNA probes in cytogenetic FISH techniques is reviewed and applications of PNA specific for non-telomeric DNA are illustrated.

*Correspondence to: Dr. Peter M. Lansdorp, Terry Fox Laboratory, British Columbia Cancer Research Centre, 675 West 10th Avenue, Vancouver, BC, V5Z 1L3, Canada. Phone: 604-675 8135 ; Fax: 604-877-0712 ; E-mail: plansdor@bccrc.ca

Keywords: quantitative fluorescence in situ hybridization; peptide nucleic acid probes; telomere repeats; repetitive DNA sequences, metaphase chromosomes

Introduction

Recombinant DNA technology, which allowed production of large quantities of known DNA or RNA sequences in bacteria, paved the way for in situ hybridization methods to map mammalian genes and study gene expression in cells and tissue sections in the 1970's. The first fluorescent in situ hybridization (FISH) experiments used mRNA molecules which were linked to fluorescent latex microspheres to visualize globin genes in human metaphase chromosomes [1]. Since then, direct labeling of probes with suitable fluorescent dyes was introduced and many additional improvements in FISH methods and fluorescence detection by fluorescence microscopy were made (reviewed in [2]). As a result, FISH became a valuable *qualitative* tool in cytogenetics and gene mapping.

In the late eighties, the first reports can be found describing in situ hybridization of probes to interphase nuclei in suspension followed by the quantification of bound probe by flow cytometry [3]. The aim in those studies was to combine FISH with the speed and quantitative analysis provided by flow cytometry. Improvements in FISH probes, antibody labeling, fluorescence labels and flow cytometers all contributed to further advances in FISH techniques using either fluorescence microscopy or flow cytometry. However, one of the fundamental drawbacks of conventional FISH techniques aimed to detect targets sequences in human chromosomes remained. This limitation is that the DNA or RNA probes have to compete with complementary DNA strands for hybridization to the denatured target sequences during the hybridization process. While labeling both strands of probes and high probe concentrations will increase the likelihood of specific hybridization, this does not solve the basic problem especially if short oligonucleotide probes specific for repeat sequences have to compete with long strands of complementary DNA. This limitation of conventional FISH is illustrated by the use of biotinylated $(GGGTTA)_7$ and $(TAACCC)_7$ deoxynucleotide oligomeric probes to detect telomeric repeats in human metaphase chromosomes [4]. In these studies signals were detected at some but not all chromosome ends. Limitations imposed by the negatively charged phosphate backbone of conventional DNA or RNA oligonucleotide probes in FISH techniques were predicted to be less of a concern with a new sort of synthetic "oligomimetic" molecule in which nucleotides are attached to an uncharged "peptide-like" backbone: Peptide Nucleic Acids (PNA, reviewed in [5]). This prediction was confirmed when it was found that directly fluorescently labeled PNA probes could be hybridized to target sequences using conditions that *did not allow* re-annealing of complementary strands [6]. These observations allowed development of the first *quantitative* FISH (Q-FISH) techniques.

Peptide Nucleic Acids (PNA)

Peptide nucleic acid (PNA) is a nucleic acid analog in which the sugar phosphate backbone of natural nucleic acid has been replaced by a synthetic peptide backbone formed from N-(2-amino-ethyl)-glycine units, resulting in an achiral and uncharged moiety that mimics RNA or DNA oligonucleotides [5]. PNA is chemically stable and resistant to hydrolytic (enzymatic) cleavage and cannot be degraded inside living cells. PNA is capable of sequence-specific recognition of DNA and RNA obeying the Watson-Crick hydrogen bonding scheme, and the hybrid complexes exhibit extraordinary thermal stability and unique ionic strength properties (reviewed in [7]). PNA resulted from the work by Peter Nielsen and Michael Egholm in Ole Buchardt's laboratory in Denmark. The unique properties of PNA-DNA hybrid complexes were exploited in FISH techniques in 1995 by Peter Lansdorp during a sabbatical visit in the laboratory of Hans Tanke and Ton Raap at Leiden University in the Netherlands. While directly labeled PNA probes were already commercially available, their use in FISH techniques required the development of conditions that allowed efficient hybridization at low ionic strength (preventing competition between probe hybridization and DNA renaturation) and low background staining of the labeled probe. The latter was particularly problematic because PNA probes are very hydrophobic and strongly bind e.g. to glass and various non-specific (hydrophobic) sites in cells and chromosomes. Solutions to these problems were found by including a blocking protein solution in the hybridization mixture and performing hybridization and wash steps in the presence of high concentrations (70%) of formamide [6]. As in all protocol development efforts, countless experiments had to be performed to delineate the importance of each individual parameter in the technique, identify crucial variables and establish a procedure that works reproducibly. Details of the protocols that are in current use are discussed elsewhere [8, 9]. Once "clean" and bright signals at all telomeres were obtained it was observed that 1) the telomere spots at sister chromatids are typically of very similar intensity, in line with the proposed quantitative nature of the PNA- FISH hybridization and 2) the telomere length at individual chromosomes is remarkably variable [6]. This variation in repeat length at any given chromosome end was also observed in murine metaphase chromosomes [10]. By combining karyotype and Q-FISH analysis it was found that, despite this large variation, the *average* telomere length at specific chromosome arms shows reproducible differences between specific chromosomes and between individuals [11]. For example, the short arm of chromosome 17p was found to typically contain fewer telomere repeats than most other chromosome arms. While these finding have been confirmed by different groups [12-14], the factors that determine the differences in chromosome-specific telomere length remain elusive. The observation that the short p arms of acrocentric mouse chromosomes (adjacent to centromeric DNA) are typically have fewer telomere repeats than the long q arms [10] have fueled speculation that the presence of specific sequences and/or associated heterochromatin is a major factor determining telomere length [15]. One possibility is that, with heterochromatin nearby, telomeric DNA is replicated late during the cell cycle allowing perhaps less time for elongation by telomerase [16]. Alternatively, repair of telomeric DNA may be different as a result of variation in the overall "chromatin" environment. The latter could possibly explain the observation that telomere repeats in inactive X chromosomes in female cells are lost at an accelerated rate with age as compared to active X chromosomes [16]. Recently it was suggested that elongation by telomerase is required for repair of damage to telomeric DNA [17] and such repair could be compromised at inactive X chromosomes. Given the large variation in telomere length between individual chromosome arms [6, 10, 11, 18] and the observed differences in the rate of telomere loss between inactive and active X chromosomes [16], it is perhaps somewhat surprising that the average telomere length measured by Q-FISH is nevertheless strongly correlated with the involvement of specific chromosome arms in genetic instability in various model systems [19-22]. These observations suggest that further evaluation of the (heritable) differences in average telomere length at specific chromosome arms as risk factors for the development of specific types of cancer and other disorders is warranted. Unfortunately, such studies are not only very time-consuming using Q-FISH techniques but also complicated by allele-specific differences between specific parental chromosome arms [12, 13]. Allele-specific PCR techniques such as the STELA techniques developed by Baird et al [23] could possibly be used to further asses the role of chromosome-specific telomere length as risk factors in the development of specific disorders. Unfortunately, suitable PCR primers specific for chromosome and allele-specific sub-telomeric sequences have been identified for only a very limited number of human chromosome arms.

Conclusion

PNA probes are very specific for complementary single strand DNA or RNA and poorly tolerate base-pair mismatches between probe and target sequences [5, 24]. Unlike typical genomic sequences telomere repeats are maintained by telomerase, a specialized reverse transcriptase that adds TTAGGG repeats to the 3' end of chromosomes using a dedicated RNA template. This mechanism ensures that at least the very termini of chromosome ends will be characterized by invariant repeats that provide an ideal target for specific PNA probes. The frequency of perfect matches between probe and target sequence is typically much lower for other than telomere repeats in mammalian genomes. Sequence variation between genomic repeats will limit the signal obtained with PNA probes. However because certain repeats (e.g. Alu repeats or certain satellite sequences at centromeres) are much more abundant than telomere repeats, single probes can nevertheless provide strong and specific signals [19, 25]. This illustrated in Figure 1 showing the result of PNA FISH on a human metaphase cell using multiple PNA probes each labeled with a different fluorochrome. Next to very clear signals at chromosome ends (Figure 1A), strong signals are obtained using a fluorescein labed PNA probe specific for alpha satellite DNA (Figure 1B) and a Cy5 labeled probe specific for human Alu repeats (Figure 1C). For less abundant repeats (e.g. chromosome specific repeats) it may be necessary to combine multiple probes [26]. The selection of suitable probes can be cumbersome because, despite the well advertised "completion of the human genome sequence" several years ago [27, 28], information about repetitive centromeric DNA is still far from complete even in the most current genome databases.

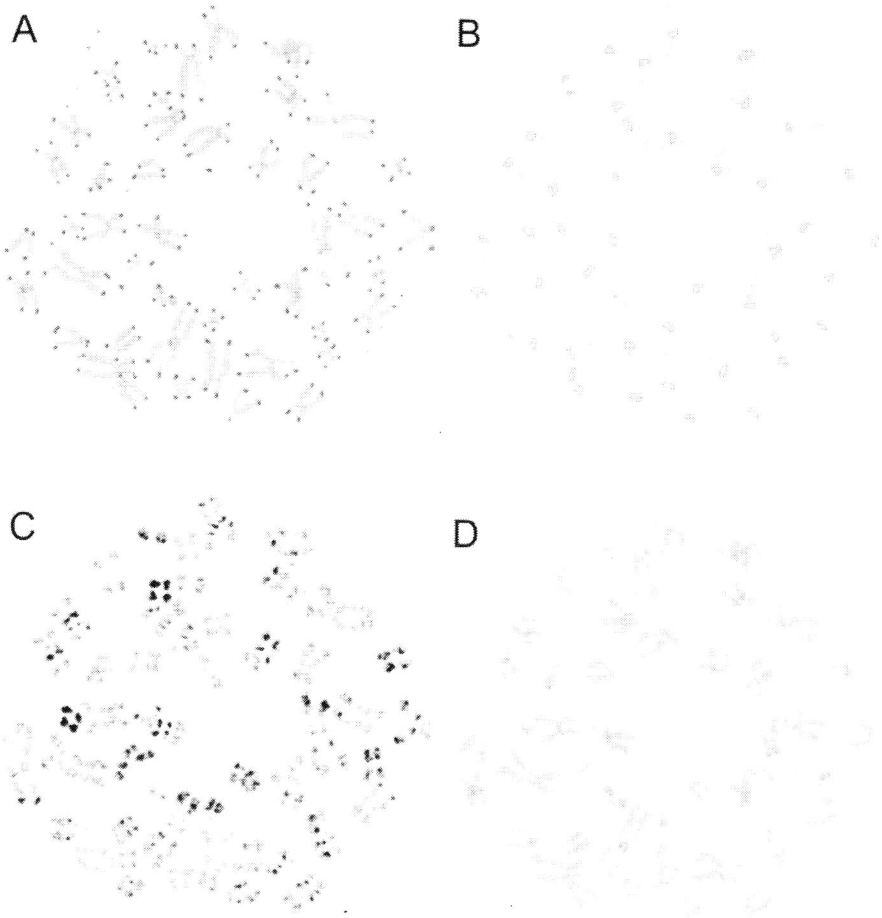

Figure 1. PNA probes are powerful tools to detect repeat sequences in chromosomes. In this experiment a normal human lymphocyte metaphase was hybridized with a mixture of three different directly labeled PNA probes: a Cy3 labeled $(CCC\ TAA)_3$ probe specific for telomeric repeats, a FITC labeled CTT CGT TGG AAA CGG GGT probe specific for centromeric alpha satellite DNA and a Cy5 labeled CCT CAG CCT CCC OOO JJJ TJJ probe specific for Alu repeats. The latter probe targets GGG AGG CTG AGG sequences in Alu repeats. The O residues have no nucleotide specificity and serve as spacers to allow the J and T residues in the distal part of the probe to fold back and form a triplex with the GGG AGG part of the Alu sequence. Following FISH chromosomes were counterstained with DAPI. Images at four different wavelengths were acquired with a 63x 1.4 oil objective over a 4 micron focal distance at 0.15 micron intervals using a DeltaVision optical sectioning microscope. Shown are pseudo-colored deconvolved images of DAPI only (D), probe only (C) or projections obtained with the individual PNA probe superimposed on the image obtained with DAPI (A and B)

Unpredicted cross hybridization, variable background staining of individual PNA probes and the expense of PNA probes continue to pose a serious barrier to the application of mixtures of PNA probes to detect of unique DNA sequences. Thus despite all the advantages of PNA that were mentioned in this mini-review, it seems possible that the very first application of PNA probes in human cytogenetic analysis, the quantitative detection of telomere repeat sequences [6], will remain the most widely used application of these for powerful synthetic compounds.

Acknowledgements

Liz Chavez is thanked for performing the experiment shown in Figure 1 and Steven Poon for discussion and help with image analysis. Applied Biosystems and Jim Coull are thanked for making labeled PNA probes available for these studies. The work in the laboratory of the author is supported by grants from the National Institutes of Health (AI29524), the Canadian Institute of Health Research (MOP38075) and the National Cancer Institute of Canada with funds derived from the Terry Fox Run.

References

[1] Cheung, S.W., et al., Gene mapping by fluorescent in situ hybridization. *Cell Biol Int Rep*, 1977. 1(3): p. 255-62.

[2] Trask, B.J., Fluorescence in situ hybridization: applications in cytogenetics and gene mapping. *Trends Genet*, 1991. 7(5): p. 149-54.

[3] Trask, B., et al., Fluorescence in situ hybridization to interphase cell nuclei in suspension allows flow cytometric analysis of chromosome content and microscopic analysis of nuclear organization. *Hum Genet*, 1988. 78(3): p. 251-9.

[4] Moyzis, R.K., et al., A highly conserved repetitive DNA sequence, (TTAGGG)n, present at the telomeres of human chromosomes. *Proc Natl Acad Sci U S A,* 1988. 85(18): p. 6622-6.

[5] Nielsen, P.E. and M. Egholm, An introduction to peptide nucleic acid. *Curr Issues Mol Biol,* 1999. 1(1-2): p. 89-104.

[6] Lansdorp, P.M., et al., Heterogeneity in telomere length of human chromosomes. *Hum Mol Genet,* 1996. 5(5): p. 685-91.

[7] Ray, A. and B. Norden, Peptide nucleic acid (PNA): its medical and biotechnical applications and promise for the future. *Faseb J,* 2000. 14(9): p. 1041-60.

[8] Poon, S.S. and P.M. Lansdorp, Measurements of telomere length on individual chromosomes by image cytometry. *Methods Cell Biol,* 2001. 64: p. 69-96.

[9] Poon, S.S., et al., Telomere length measurements using digital fluorescence microscopy. *Cytometry,* 1999. 36(4): p. 267-78.

[10] Zijlmans, J.M., et al., Telomeres in the mouse have large inter-chromosomal variations in the number of T2AG3 repeats. *Proc Natl Acad Sci U S A*, 1997. 94(14): p. 7423-8.

[11] Martens, U.M., et al., Short telomeres on human chromosome 17p. *Nat Genet,* 1998. 18(1): p. 76-80.

[12] Graakjaer, J., et al., The pattern of chromosome-specific variations in telomere length in humans is determined by inherited, telomere-near factors and is maintained throughout life. *Mech Ageing Dev,* 2003. 124(5): p. 629-40.

[13] Graakjaer, J., et al., The relative lengths of individual telomeres are defined in the zygote and strictly maintained during life. *Aging Cell,* 2004. 3(3): p. 97-102.

[14] Leach, N.T., et al., Human chromosomes with shorter telomeres and large heterochromatin regions have a higher frequency of acquired somatic cell aneuploidy. *Mech Ageing Dev,* 2004. 125(8): p. 563-73.

[15] Slijepcevic, P., Telomere length and telomere-centromere relationships? *Mutat Res*, 1998. 404(1-2): p. 215-20.

[16] Surralles, J., et al., Accelerated telomere shortening in the human inactive X chromosome. *Am J Hum Genet,* 1999. 65(6): p. 1617-22.

[17] Lansdorp, P.M., Major cutbacks at chromosome ends. *Trends Biochem Sci,* 2005. 30(7): p. 388-95.

[18] Martens, U.M., et al., Accumulation of short telomeres in human fibroblasts prior to replicative senescence. *Exp Cell Res*, 2000. 256(1): p. 291-9.

[19] Deng, W., et al., Distinct profiles of critically short telomeres are a key determinant of different chromosome aberrations in immortalized human cells: whole-genome evidence from multiple cell lines. *Oncogene,* 2004. 23(56): p. 9090-101.

[20] Gisselsson, D., et al., Telomere-mediated mitotic disturbances in immortalized ovarian epithelial cells reproduce chromosomal losses and breakpoints from ovarian carcinoma. *Genes Chromosomes Cancer,* 2005. 42(1): p. 22-33.

[21] Hemann, M.T., et al., The shortest telomere, not average telomere length, is critical for cell viability and chromosome stability. *Cell*, 2001. 107(1): p. 67-77.

[22] Zou, Y., et al., Does a sentinel or a subset of short telomeres determine replicative senescence? *Mol Biol Cell*, 2004. 15(8): p. 3709-18.

[23] Baird, D.M., et al., Extensive allelic variation and ultrashort telomeres in senescent human cells. *Nat Genet*, 2003. 33(2): p. 203-7.

[24] Egholm, M., et al., PNA hybridizes to complementary oligonucleotides obeying the Watson-Crick hydrogen-bonding rules. *Nature*, 1993. 365(6446): p. 566-8.

[25] Chen, C., et al., Single base discrimination of CENP-B repeats on mouse and human Chromosomes with PNA-FISH. *Mamm Genome*, 1999. 10(1): p. 13-8.

[26] Taneja, K.L., et al., Multicolor fluorescence in situ hybridization with peptide nucleic acid probes for enumeration of specific chromosomes in human cells. Genes *Chromosomes Cancer,* 2001. 30(1): p. 57-63.

[27] Lander, E.S., et al., Initial sequencing and analysis of the human genome. *Nature*, 2001. 409(6822): p. 860-921.

[28] Venter, J.C., et al., The sequence of the human genome. *Science,* 2001. 291(5507): p. 1304-51.

In: PRINS and PNA Technologies in Chromosomal Investigations ISBN 1-60021-413-4
Editor: Franck Pellestor, pp. 177-183

Chapter 12

PNA Technique for *In Situ* Assessment of Aneuploidy in Human Sperm

***Franck Pellestor*[*,1,2,3], *Brigitte Andréo* [2] *and Samir Hamamah* [1,3]**

[1] Department of Reproductive Biology B,
Arnaud de Villeneuve Hospital, Montpellier, France
[2] Institute of Human Genetics, CNRS UPR 1142, Montpellier, France
[3]Inserm, U 847, Montpellier, France

Abstract

The Peptide Nucleic Acids (PNAs) in situ labelling techniques constitute an attractive alternative to FISH and PRINS procedures for chromosomal in situ investigations. PNA probes are synthetic DNA analogs with uncharged polyamide backbones. The PNA labelling reaction presents several advantages (specificity, rapidity, discriminating ability, short size of PNA probes) that make it very attractive for cytogenetic purposes. The efficiency of centromeric PNA probes has recently been demonstrated on human cell preparations. Their adaptation on human spermatozoa has allowed the development of a new and fast procedure, which can advantageously be used to complement FISH and PRINS techniques for the in situ assessment of aneuploidy in male gametes.

Key-words: Aneuploidy; Chromosomal screening; PNA; Spermatozoa

[*] Address for correspondence: Dr. Franck Pellestor, Department of Reproductive Biology B, Hôpital Arnaud de Villeneuve 371, avenue du Doyen Gaston Giraud, 34295 Montpellier Cedex 5, France. Tel: 33 (0)4 67 33 64 04; E-mail: Franck.Pellestor@igh.cnrs.fr

Introduction

Peptide Nucleic Acids (PNAs) constitute a new class of DNA probes, which provide an interesting complement to FISH and PRINS in the field of human cytogenetics. PNAs are synthetic mimics of DNA in which the deoxyribose phosphate backbone supporting the nucleic acid bases is replaced by a non-charged peptide backbone. The unique chemical makeup of these molecules confers a number of beneficial properties, including enhanced hybridisation rates, resistance to nucleases and proteases and the ability to penetrate condensed biological structures [1]. The neutral backbone of PNA provides strong binding between PNA/DNA or PNA/RNA strands and greater specificity of interaction than their DNA counterparts. While they hybridise according to normal Watson-Crick base pairing rules, PNA have been shown to bind to DNA or RNA targets with higher affinity than the corresponding oligonucleotides. This high level of discrimination at single-base level has indicated that short PNA probes could offer high specificity and has thus allowed the further development of several PNA-based strategies for molecular investigations and diagnosis [2]. Short PNA oligomers, from17 to 22 bases units, constitute efficient tools for detecting specific DNA sequences with fast hybridisation kinetics.

The unique properties of PNA probes as a DNA mimic have led to the development of numerous applications. Most notably, PNAs find current uses in molecular biological techniques as specific and sensitive probes for complementary nucleic acids [3]. The introduction of PNA technology in cytogenetics is recent. Firstly, it has been demonstrated that PNA probes were useful for detecting telomere repeat sequences [4,5]. Then, the availability of chromosome-specific centromeric PNA probes, directly fluorochrome-labelled, has led to the development of rapid and easy multi-colour PNA protocols for the in situ detection and enumeration of human chromosomes in metaphases and interphase nuclei [6]. The procedure has been recently adapted to human blastomeres, and gametes [7, 8] and opens news and promising perspectives for PNA in the field of genetic diagnosis.

Application of PNA on Human Sperm

Direct chromosomal analysis of spermatozoa is an essential approach for the investigation of the occurrence and aetiology of chromosomal abnormalities in humans under a wide variety of clinical conditions. To date, numerous chromosomal analyses on human sperm have been performed using fluorescence in situ hybridization (FISH) [9,10]. These reports demonstrated the efficiency of the in situ labelling procedure on male gametes, but also pointed out the limitations of FISH on this biological material, which are essentially linked to the size of the probes and the reliability of the associated sperm decondensation treatments [11]. The primed in situ (PRINS) reaction has offered an alternative approach for the direct chromosome analysis of human spermatozoa [12]. Several studies have demonstrated that PRINS labelling was as efficient as FISH on human sperm and exhibited higher specificity, but its current use requires sequential reactions [13, 14].

PNA provide a new alternative to FISH and PRINS for sperm nuclei analysis. The adaptation of this new class of probes to human spermatozoa constituted a new step in the development of PNA methodology and a very interesting challenge because of the particularities of human sperm nucleus in terms of genomic compaction and accessibility of DNA sequences.

When applied to human sperm preparation, FISH and PRINS labelling techniques need to be associated with a sperm nucleus decondensing treatment in order to make the targeted DNA region accessible. The quality of the fluorescent signals, and consequently the reliability of the sperm analysis results are directly related to the efficiency of the in situ decondensation [15, 16]. Pretreatment to decondense sperm chromatin is also indispensable with PNA probes. Indeed, no PNA labelling of sperm nuclei is obtained without decondensation pretreatment. Consequently, we have tested the 2 decondensation procedures which give efficient and reproducible results with FISH and PRINS reactions, i.e. the DTT and the NaOH treatments. The two procedures yielded satisfactory results with PNA (Fig. 1) and gave similar kinetics for the labelling reaction.

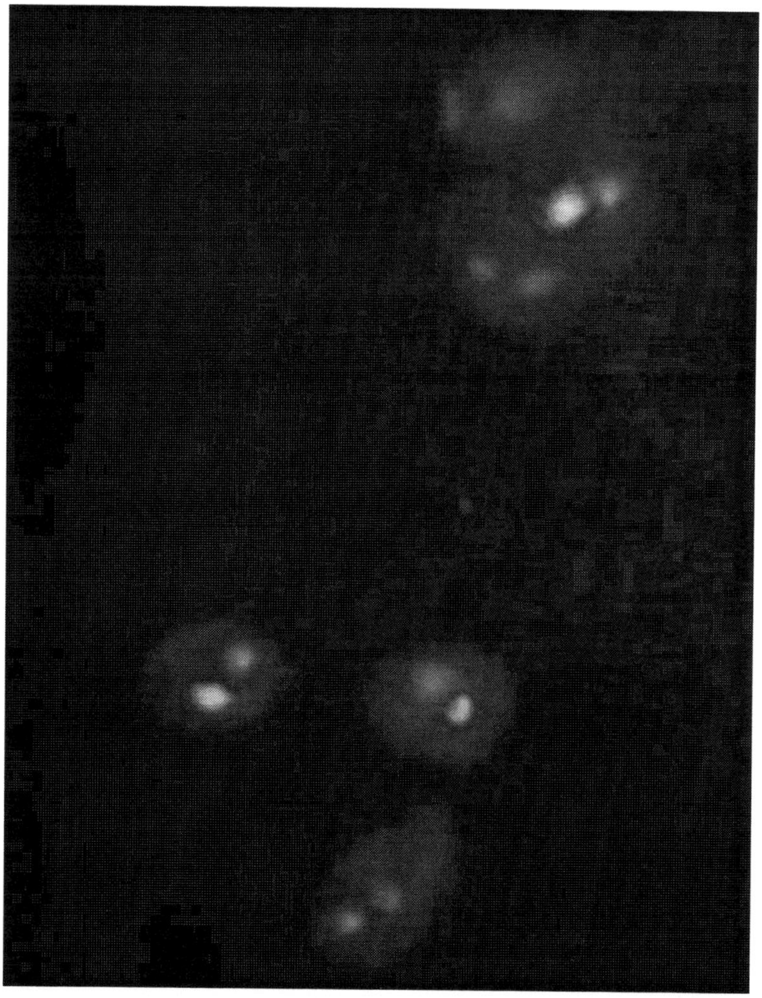

Figure 1. Example of PNA labelling on human sperm nuclei using centromeric chromosome-specific PNA probes. The chromosome 1 is labelled in blue, the chromosome X in red and the chromosome Y in green.

In order to assess the limit of PNA hybridization, several durations of hybridization time (from 5 to 60 minutes) were tested. Satisfactory results, i.e. well-distinguishable spots, were obtained in sperm preparations following 30 – 60 minutes hybridization. Shortest hybridizations did not allow enough time to obtain an efficient labelling of the targeted chromosomes. The hybridization timing of PNA probes appeared to be considerably shortened in comparison to FISH reaction on sperm, which requires an overnight hybridization in order to be efficiently completed. The fast hybridization kinetics of PNA is more similar to the kinetics of PRINS reaction on which sperm chromosomal labelling can be performed in less than 30 minutes [13]. The similarity between PNA and PRINS might be essentially due to the small size of both PNA oligoprobes (18-20 bp) and PRINS primers, which do not exceed 30 bases in length. Reduced labelling timing (around three hours) has also been reported in some sperm FISH studies, but using short satellite oligonucleotide probes [17, 18]. This data points out the importance of the probe size, and the use of a decondensation procedure in sperm labelling assays.

Similar performance of PNA and PRINS methods has already been reported for the in situ detection and sizing of telomeric repeat sequences [19]. Both techniques presented comparable features in terms of specificity, staining intensity and efficiency, but PRINS always displayed faster turnaround reaction time. This could reflect the fact that PRINS is an "active" reaction involving an ultra-fast biochemical reaction of primer extension catalysed by a Taq polymerase. In the case of PNA reaction, the rapidity of the labelling is due to the neutral backbone of the PNA molecules, which allows for PNA/DNA binding to occur more rapidly and more tightly than DNA/DNA binding. Moreover, PNA probes can bind to DNA under low ionic strength conditions that disfavour reannealing of complimentary genomic strands. This advantage is particularly important for in situ hybridization experiments. The high affinity of PNA probes to DNA constitutes an important feature for chromosomal analysis on human gametes. Previous studies have demonstrated that PNA probes could discriminate between two centromeric DNA repeats that differ by only a single base pair [20], whereas standard FISH probes are unable to discriminate sequences with a single base resolution. The study of chromosomal polymorphism could benefit from the discriminating power of PNA probes. On sperm, this could be useful for the in situ distinction of autosomal nondisjunction occurring at meiosis I and meiosis II, when satellite polymorphisms exist between two homologous chromosomes.

To estimate and validate the efficiency of PNA labelling on human sperm, comparative estimates of disomy for chromosomes X, Y and 1 were performed on sperm preparations from healthy subjects using multicolour FISH, PRINS and PNA procedures in parallel. An equivalent quality of in situ nuclear labelling and similar disomy rates were obtained with the 3 methods (Table 1). These results were in agreement with the data previously reported in the literature.

Table 1. Results of sex ratio and disomy estimates for gonosomes and chromosome 1 using PNA, PRINS and FISH techniques on sperm preparations from 2 normal men

Methodology	Donor	N. of screened nuclei	Sex ratio		Overall sex chromosome disomy (%)	Disomy			
			X	Y	XX + YY + XY	XX	YY	XY	chr. 1
PNA	1	10008	4981 (49.77)	4990 (49.86)	0.36	16 (0.14)	10 (0.10)	12 (0.12)	13 (0.13)
	2	10036	5010 (49.92)	4998 (49.80)	0.28	8 (0.08)	10 (0.10)	10 (0.10)	18 (0.18)
PRINS	1	10040	4999 (49.79)	4992 (49.72)	0.39	9 (0.09)	13 (0.13)	17 (0.17)	10 (0.10)
	2	10005	4988 (49.85)	4981 (49.78)	0.36	12 (0.12)	14 (0.14)	10 (0.10)	23 (0.23)
FISH	1	10010	4987 (49.82)	5023 (50.17)	0.40	11 (0.11)	10 (0.10)	19 (0.19)	13 (0.13)
	2	10013	5009 (50.03)	5004 (49.97)	0.32	9 (0.09)	10 (0.10)	13 (0.13)	21 (0.21)

Conclusion

Since its introduction, the PNA technique has quickly evolved from basic research to diagnosis procedures. The fast hybridisation, penetration and discrimination of PNA probes make them valuable tools for *in situ* chromosomal screening. Their successful use on human sperm has proven that these reagents could be used on difficult biological material, and consequently they have a great potential for clinical application. With the development of rapid and simplified protocols producing reliable and reproducible results, the PNA technique has proven that it could become powerful tool for cytogenetic investigations and diagnosis. The PNA method can advantageously be used to complement FISH and PCR for the physical mapping of the human genome, and the possibility of combining PNA with FISH or PRINS reactions on a same preparation opens up interesting possibilities for multiplex assays.

Acknowledgments

This study was supported by a European INTAS project (N° 03-51-4060).

References

[1] Nielsen, P.E., and Egholm, M. (1999) An introduction to peptide nucleic acid. *Curr. Issues Mol. Biol.* 1, 89-104.

[2] Ray, A., and Norden, B. (2000) Peptide nucleic acid (PNA): its medical and biotechnical applications and promise for the future. *FASEB J.* 14 , 1041-60.

[3] Nielsen, P.E. (2001) Peptide nucleic acid: a versatile tool in genetic diagnostics and molecular biology. *Curr. Opin. Biotechnol.* 12, 16-20.

[4] Lansdorp, P.M., Verwoerd, N.P., van de Rijke, F.M., Dragowska, V., Little, M.T., Dirks, R.W., Raap, A.K., and Tanke, H.J. (1996) Heterogeneity in telomere length of human chromosomes. *Hum. Mol. Genet.* 5, 685-691.

[5] Boei, J.J.W.A., Vermeulen, S., and Natarajan, A.T. (2000) Analysis of radiation-induced chromosomal aberrations using telomeric and centromeric PNA probes. *Int. J. Rad. Biol.* 76, 163-167.

[6] Taneja, K.L., Chavez, E.A., Coull, J., and Lansdorp, P.M. (2001) Multicolor fluorescence in situ hybridization with peptide nucleic acid probes for enumeration of specific chromosomes in human cells. *Gen. Chrom. Cancer* 30, 57-63.

[7] Pellestor, F., Andréo, B., Taneja, K., and Williams, B. (2003) PNA on human sperm: a new approach for in situ aneuploidy estimation. *Eu. J. Hum. Genet.* 11, 337-341.

[8] Paulasova, P., Andréo, B., Diblik, J., Macek, M., and Pellestor, F. (2004) The peptide nucleic acids as probes for chromosomal analysis: application to human oocytes, polar bodies and preimplantation embryos. *Mol. Hum. Reprod.* 10, 467-472.

[9] Downie, S.E., Flaherty, S.P., and Matthews, C.D. (1997) Detection of chromosomes and estimation of aneuploidy in human spermatozoa using fluorescence in-situ hybridization. *Mol. Hum. Reprod.* 3, 585-598.

[10] Martin, R.H., Rademaker, A.W., Hildebrand, K., Long-Simpson, L., Peterson, D., and Yamamoto, J. (1987) Variation in the frequency and type of sperm chromosomal abnormalities among normal men. *Hum. Genet.* 77, 108-114.

[11] Egozcue, J., Blanco, J., and Vidal, F. (1997) Chromosome studies in human sperm nuclei using fluorescence in situ hybridization (FISH). *Hum. Reprod. Update* 5, 441-452.

[12] Pellestor, F., Girardet, A., Lefort, G., Andréo, B. and Charlieu, J.P. (1995) PRINS as a method for rapid chromosomal labeling on human spermatozoa. *Mol. Reprod. Dev.* 40, 333-337.

[13] Pellestor, F., Girardet, A., Coignet, L., Andréo, B., and Charlieu, J.P. (1996) Assessment of aneuploidy for chromosomes 8, 9, 13, 16 and 21 in human sperm by using primed in situ labeling technique. *Am. J. Hum. Genet.* 58, 797-802.

[14] Pellestor, F., Imbert, I., and Andréo, B. (2002) Rapid chromosome detection by PRINS in human sperm. *Am. J. Med. Genet.* 107, 109-114.

[15] Wyrobek, A.J. (1993) Methods and concepts in detecting abnormal reproductive outcomes of paternal origin. *Reprod. Toxicol.*7, 3-16.

[16] Martini, E., Speel, E.J., Geraedts, J.P., Ramaekers, F.C., and Hopman, A.H. (1995) Application of different in-situ hybridization detection methods for human sperm analysis. *Hum. Reprod.* 10, 855-861.

[17] O'Keefe, C.L., Warburton, P.E., and Matera, A.G. (1996) Oligonucleotide probes for alpha satellite DNA variants can distinguish homologous chromosomes by FISH. *Hum. Mol. Genet.* 5,1793-1739.

[18] O'Keefe CL, Griffin DK, Bean CJ, Matera AG, Hassold TJ. (1997) Alphoid variant-specific FISH probes can distinguish autosomal meiosis I from meiosis II non-disjunction in human sperm. *Hum. Genet.* 101, 61-66.

[19] Serakinci, N., and Koch, J. (1999) detection and sizing of telomeric repeat DNA in situ. *Nat. Biotechnol.*17, 200-201.

[20] Chen, C., Hong, Y.K., Ontiveros, S.D., Egholm, M., and Strauss, W.M. (1999) Single base discrimination of CENP-B repeats on mouse and human chromosomes with PNA-FISH. *Mamm. Genome* 10,13-18.

Index

A

B

C

D

E

F

G

H

I

J

K

L

N

O

P

Q

R

S

T

U

V

W

X

Y

Z